# Groundwater Resources and Salt Water Intrusion in a Changing Environment

# Groundwater Resources and Salt Water Intrusion in a Changing Environment

Special Issue Editors

**Maurizio Polemio**
**Kristine Walraevens**

MDPI • Basel • Beijing • Wuhan • Barcelona • Belgrade

**MDPI**

*Special Issue Editors*
Maurizio Polemio
Italian National Research Council-Research
Institute for Geo-Hydrological Protection
Italy

Kristine Walraevens
Ghent University
Belgium

*Editorial Office*
MDPI
St. Alban-Anlage 66
4052 Basel, Switzerland

This is a reprint of articles from the Special Issue published online in the open access journal *Water* (ISSN 2073-4441) from 2017 to 2019 (available at: https://www.mdpi.com/journal/water/special_issues/salt_water_intrusion)

For citation purposes, cite each article independently as indicated on the article page online and as indicated below:

LastName, A.A.; LastName, B.B.; LastName, C.C. Article Title. *Journal Name* **Year**, *Article Number*, Page Range.

**ISBN 978-3-03921-197-5 (Pbk)**
**ISBN 978-3-03921-198-2 (PDF)**

Cover image courtesy of Maurizio Polemio.

# Contents

# About the Special Issue Editors

**Maurizio Polemio** was born in 1963, and graduated with a Master's degree (with honours) in Civil Engineering (Hydraulic Section) from the University of Bari, Italy, in 1987. He joined the Bari Department of the Research Institute for Hydrogeological Protection (IRPI) of the National Research Council (CNR) in 1989 as a Research Assistant. He has been a researcher at CNR since 1992, and was in charge of the Bari Department of IRPI from 2007 to 2013. He is currently the head of the Hydrology Laboratory and the Hydrogeological Research Group of CNR-IRPI. He is the author of more than 220 papers, more than 100 of which have been published in scientific journals, the majority being of a prestigious international level (https://orcid.org/0000-0002-0343-5339). Concerning the bibliometric indicators related to publications and citations, a Scopus descriptive statistic could be: 65 total documents, 658 total citations by 479 documents, H-index equal to 15 (22 for Google-scholar database), 150 coauthors. Over the years, he has attended, lectured at or organised many international and national conferences, workshop and courses to present updates of his activity, exchange views and gauge his achievements. He has a wealth of editorial experience; at present, he is a member of the Scientific Committee of Acque Sotterranee, a leading Italian hydrogeological journal. He has edited or coedited several scientific books. He has been part of several research projects, very often as the scientific lead of the project. He was the person in charge of the scientific–technical International Hydrological Programme (IHP) Secretariat of UNESCO from 2002 to 2017. He has been entrusted with university teaching as a Temporary Professor of the University of Calabria of Engineering Geology and Applied Hydrogeology and a Temporary Professor of the University of Bari of Engineering Geology. He has been the supervisor of multiple degrees and theses. He was approved by national selection as a Full Professor of Engineering Geology. He has great expertise across practical and hydrogeology, hydrology and engineering geology aspects of studied phenomena and in pursing their quantitative characterisation, on major topics including: hydrological and hydrogeological in situ measurements, surveys and monitoring, especially in the case of coastal carbonate aquifers; hydrogeological conceptualisation; water resources management in Mediterranean regions, quality degradation and vulnerability of groundwater; quantity degradation of groundwater resources due to overexploitation or to climatic change; hydrogeological characterisation and monitoring of aquifers and slopes; numerical modelling of groundwater flow and transport, and also for slope stability analysis; geostatistics and hydrological statistics of hydrogeological time series; relationship between rainfall and recurrence of damaging hydrogeological events as landslides and floods; protection against impacts of climate change and adaptation solutions; coastal aquifers and salt water intrusion; and groundwater resources and management.

**Kristine Walraevens** is a Professor of Hydrogeology at Ghent University, Belgium. She is leading the Laboratory for Applied Geology and Hydrogeology at the Department of Geology. Her research is mainly related to regional aquifers and to groundwater chemistry, including salt water intrusion in coastal aquifers, the impact of groundwater exploitation on quantity and quality of groundwater, groundwater recharge, groundwater type evolution and residence times, and nitrate pollution of groundwater. Several aspects of groundwater policy in Flanders have been based on her research. Her regional focus is on Flanders, Belgium, and on Africa, particularly Eastern Africa.

*water*  MDPI

*Editorial*

# Recent Research Results on Groundwater Resources and Saltwater Intrusion in a Changing Environment

**Maurizio Polemio** [1,*] **and Kristine Walraevens** [2]

[1]   CNR-IRPI, National Research Council-Research Institute for Geo-Hydrological Protection,
      Via Amendola 122/I, 70126 Bari, Italy
[2]   Laboratory for Applied Geology and Hydrogeology, Ghent University, Krijgslaan 281-S8, 9000 Gent,
      Belgium; kristine.walraevens@ugent.be
*    Correspondence: m.polemio@ba.irpi.cnr.it

Received: 15 May 2019; Accepted: 19 May 2019; Published: 29 May 2019

**Abstract:** This Special Issue presents the work of 30 scientists of 11 countries. It confirms that the impacts of global change, resulting from both climate change and increasing anthropogenic pressure, are huge on worldwide coastal areas (and very particularly on some islands of the Pacific Ocean), with highly negative effects on coastal groundwater resources, widely affected by seawater intrusion. Some improved research methods are proposed in the contributions: using innovative hydrogeological, geophysical, and geochemical monitoring; assessing impacts of the changing environment on the coastal groundwater resources in terms of quantity and quality; and using modelling, especially to improve management approaches. The scientific research needed to face these challenges must continue to be deployed by different approaches based on the monitoring, modeling, and management of groundwater resources. Novel and more efficient methods must be developed to keep up with the accelerating pace of global change.

**Keywords:** saltwater intrusion; groundwater resources; coastal aquifer; climate change; modelling; monitoring; salinization; water resources management

---

## 1. Introduction

The salinization of groundwater resources can be caused by natural phenomena and anthropogenic activities. If the global continental area of earth is considered, 16% is affected by groundwater salinization; seawater intrusion can be considered the prevalent phenomenon in terms of potential effects and risks [1]. Water and chemical fluxes, including nutrient loading, at the terrestrial/marine interface and across the sea floor provide an important linkage between terrestrial and marine environments.

Climate and global change impacts on the hydrological cycle [2], water resources, and ecosystems pose great challenges for global water and ecosystem management, especially where the ecological equilibria are strongly dependent on groundwater–surface water interaction [3]. The climate change scenarios require new and improved integrated tools for the assessment of climate change impacts on the hydrological cycle.

Coastal aquifers and ecosystems are currently under pressure globally from overexploitation and saltwater intrusion. Population growth and progressive gathering in coastal areas, climate change, and sea-level rise will increase this pressure and enhance the need for the protection and sustainable management of coastal groundwater resources and ecosystems for coastal communities in the future [4].

This Special Issue deals with hydrogeological, geophysical, and geochemical monitoring and characterization of the subsurface, involving the distribution of freshwater and saltwater; assessment of impacts resulting from the changing environment (both climate change and increasing anthropogenic pressure) on groundwater resources in coastal areas in terms of quantity and quality; and monitoring experiences and management approaches. This Special Issue presents the work of 30 scientists of

11 countries, located by the authors' place of work or study. The contributions have been grouped under three themes:

- impacts of the changing environment on the coastal groundwater resources;
- modelling of the freshwater–saltwater distribution;
- groundwater monitoring and management in coastal areas.

## 2. Impacts of the Changing Environment on Coastal Groundwater Resources

Oberle et al. [5], on the basis of monitoring data from Roi-Namur Island on Kwajalein Atoll, Marshall Islands, including electrical resistivity tomography (ERT) surveys, studied the impact of an island-overwash event, severe rainfall events, and tidal forcing of the freshwater lens on the groundwater resources of low-lying atoll islands. The overwash event was related to climate-induced local sea-level change, resulting in degradation in freshwater resources. Overwash events are likely to increase in severity in the future due to projected sea-level rises.

Stumm and Como [6] studied the saltwater intrusion using electromagnetic induction (EMI)-logging in the aquifer of southern Manhattan Island, New York. They reported that historical industrial pumping (22.7 million litres per day) early in the 20th century caused the development of several saltwater intrusion wedges. Although the pumping stopped more than 70 years ago, freshwater flow in the aquifer has been unable to push the saltwater back, due to limited recharge caused by impervious surfaces. They concluded that the glacial aquifer has had only a limited recovery from the past industrial exploitation.

Tal et al. [7] investigated the interrelationship between a multi-layered coastal aquifer at the southern Carmel plain in Israel, fish-ponds, and the sea using off-shore seismic surveying, on-land time-domain electromagnetic (TDEM) surveying, electrical conductivity (EC) profiles, hydrological field experiments, and groundwater levels. Using groundwater modelling, they showed that the exact location of the hydraulic connection between the confined aquifer unit and the sea (variable continuity of confining clay) played a significant role in the sensitivity of the aquifer unit to seawater intrusion. The geophysical methods they used helped to determine this location. They used another practical way to estimate this location, measuring the tidal amplitude in an observation well near the seashore. The authors suggested that these methods be used as managerial tools near the sea to avoid large seawater intrusion in response to impacts.

Mushtaha and Walraevens [8] quantified submarine groundwater discharge (SGD) in the Gaza Strip, Palestine, where overexploitation, seawater intrusion, and pollution by nutrients are putting the groundwater resources under high pressure. Using continuous radon measurements, they showed SGD to occur throughout the coast. High values of SGD were found in the south, and are probably related to the shallowness of the unconfined aquifer, while the lowest values of SGD were found in the middle of Gaza Strip, and they are probably related to the presence of Sabkhas. Considering that SGD would occur with the measured rates in a strip 100 m wide along the whole coast line, this results in a quantity of 38 million $m^3$ of groundwater being discharged yearly to the Mediterranean Sea along the Gaza coast. This is accompanied by a yearly discharge of over 400 tons of nitrate and 250 tons of ammonium from groundwater to the Mediterranean Sea.

## 3. Modelling of the Freshwater–Saltwater Distribution

Yoon et al. [9] used data of tide level, rainfall, groundwater level, and interface to construct time series models based on an artificial neural network (ANN) and support vector machine (SVM). Their data were for the groundwater observatory on Jeju Island, South Korea. They used the "interface egg" [10], a monitoring probe which, thanks to its specific density, can float on the freshwater–saltwater interface. They showed that the SVM-based time series model was more accurate and stable than the ANN at the study site.

Babu et al. [11] developed a methodology for regional- and well-scale modelling of an island freshwater lens based on a sharp interface approach. A quasi-three-dimensional finite element model was calibrated with freshwater thickness, where the interface was matched to the lower limit of the freshwater lens, using Tongatapu Island in the Kingdom of Tonga, a Pacific island nation, as a case study. The authors concluded that the application of a sharp interface groundwater model for real-world small islands is useful when dispersion models are challenging to implement due to insufficient data or computational resources.

Mabrouk et al. [12] assessed the situation in 2010 regarding the available fresh groundwater resources and evaluated future salinization in the Nile Delta Aquifer in Egypt, using a three-dimensional variable-density groundwater flow model coupled with salt transport with SEAWAT [13]. They examined six future scenarios that combine two driving forces: increased extraction and sea-level rise. The results showed that groundwater extraction has a greater impact on salinization of the Nile Delta Aquifer than sea-level rise, while the two factors combined cause the largest reduction of available fresh groundwater resources. The authors also determined the groundwater volumes of fresh water, brackish water, light brackish water, and saline water in the Nile Delta Aquifer. They identified the governorates that are most vulnerable to salinization.

## 4. Groundwater Monitoring and Management in Coastal Areas

Alberti et al. [14] considered the specific case of groundwater on small islands, with Nauru in the Pacific Ocean as an example, and warned for overexploitation of the thin freshwater lens and saltwater intrusion. They emphasized that the thin freshwater lens on small islands is an important resource to ensure the islands' future water security. But they emphasized that the most vulnerable aquifer systems in the world are present on small islands. Especially there, groundwater should be considered as a public and shared resource for present and future generations. The authors called for the State to directly assume the responsibility for extracting and distributing water from this vulnerable resource.

Alfarrah and Walraevens [15] studied coastal areas of arid and semi-arid regions, where the coastal aquifers are particularly at risk of saltwater intrusion, given the concentration of population along the coasts and the limited groundwater recharge. They discussed the case of Tripoli (Libya), where overexploitation has resulted in an impressive depression cone. Moreover, irrigation with nitrogen fertilizers and domestic sewage has led to high $NO_3^-$ concentration and overall pollution of the resource.

## 5. Conclusions

The increasing population density along the coasts is observed at a global scale, together with the increase in groundwater abstraction, causing problems with groundwater salinity and quantity [16]. This Special Issue confirms that the impacts of global change, resulting from both climate change and increasing anthropogenic pressure, are huge on worldwide coastal areas, with highly negative effects on coastal groundwater resources, widely affected by seawater intrusion. The well-known specific vulnerability of islands in the Pacific Ocean is clearly illustrated by the case studies presented here.

The scientific research needed to face these challenges must continue to be deployed by different approaches based on the monitoring, modelling, and management of groundwater resources. Novel and more efficient methods must be developed to keep up with the accelerating pace of global change. New surveying geophysical methods and innovative monitoring tools and equipment offer opportunities for better accuracy, higher frequency, more simplicity, and reduced costs of seawater intrusion characterisation, while new modelling solutions improve our capacity to understand groundwater systems and to predict the future effects of global change.

The further development and integration of these novel approaches is an urgent and compelling challenge. The main objectives of research should be to define optimal groundwater management criteria and to improve the performance of large-scale mathematical models to assess the impacts of

*Water* **2019**, *11*, 1118

global change on groundwater resources, using long-term up-to-date monitoring tools both to calibrate and validate modelling results.

**Funding:** This research received no external funding.

**Conflicts of Interest:** The authors declare no conflict of interest.

## References

1. IGRAC. *Global Overview of Saline Groundwater Occurrence and Genesis*; Report No. GP 2009-1; International Groundwater Resources Assessment Centre: Utrecht, The Netherlands, 2009.
2. Polemio, M.; Casarano, D. Climate change, drought and groundwater availability in southern Italy. *Geol. Soc. Spec. Publ.* **2008**, *288*, 39–51. [CrossRef]
3. De Giorgio, G.; Zuffianò, L.E.; Polemio, M. The role of the hydrogeological and anthropogenic factors on the environmental equilibrium of the Ugento Wetland (Southern Italy). *Rend. Online Della Soc. Geol. Ital.* **2019**, *47*, 79–84.
4. Langevin, C.; Sanford, W.; Polemio, M.; Povinec, P. Background and summary: A new focus on groundwater-seawater interactions. In *New Focus on Groundwater-Seawater Interactions*; Sanford, W., Langevin, C., Polemio, M., Povinec, P., Eds.; IAHS-AISH Publication: Oxford, UK, 2007; Volume 312, pp. 3–10.
5. Oberle, F.K.J.; Swarzenski, P.W.; Storlazzi, C.D. Atoll Groundwater Movement and Its Response to Climatic and Sea-Level Fluctuations. *Water* **2017**, *9*, 650. [CrossRef]
6. Stumm, F.; Como, M.D. Delineation of Salt Water Intrusion through Use of Electromagnetic-Induction Logging: A Case Study in Southern Manhattan Island, New York. *Water* **2017**, *9*, 631. [CrossRef]
7. Tal, A.; Weinstein, Y.; Wollman, S.; Goldman, M.; Yechieli, Y. The Interrelations between a Multi-Layered Coastal Aquifer, a Surface Reservoir (Fish Ponds), and the Sea. *Water* **2018**, *10*, 1426. [CrossRef]
8. Mushtaha, A.M.; Walraevens, K. Quantification of Submarine Groundwater Discharge in the Gaza Strip. *Water* **2018**, *10*, 1818. [CrossRef]
9. Yoon, H.; Kim, Y.; Ha, K.; Lee, S.-H.; Kim, G.-P. Comparative Evaluation of ANN- and SVM-Time Series Models for Predicting Freshwater-Saltwater Interface Fluctuations. *Water* **2017**, *9*, 323. [CrossRef]
10. Kim, Y.; Yoon, H.; Kim, K.P. Development of a novel method to monitor the temporal change in the location of the fresh-saltwater interface and time series models for the prediction of the interface. *Environ. Earth Sci.* **2016**, *75*, 882–891. [CrossRef]
11. Babu, R.; Park, N.; Yoon, S.; Kula, T. Sharp Interface Approach for Regional and Well Scale Modeling of Small Island Freshwater Lens: Tongatapu Island. *Water* **2018**, *10*, 1636. [CrossRef]
12. Mabrouk, M.; Jonoski, A.; Oude Essink, G.H.P.; Uhlenbrook, S. Impacts of Sea Level Rise and Groundwater Extraction Scenarios on Fresh Groundwater Resources in the Nile Delta Governorates, Egypt. *Water* **2018**, *10*, 1690. [CrossRef]
13. Langevin, C.D. *SEAWAT: A Computer Program for Simulation of Variable-Density Groundwater Flow and Multi-Species Solute and Heat Transport*; U.S. Geological Survey Fact Sheet 2009-3047; USGS: Reston, VA, USA, 2009.
14. Alberti, L.; La Licata, I.; Cantone, M. Saltwater Intrusion and Freshwater Storage in Sand Sediments along the Coastline: Hydrogeological Investigations and Groundwater Modeling of Nauru Island. *Water* **2017**, *9*, 788. [CrossRef]
15. Alfarrah, N.; Walraevens, K. Groundwater Overexploitation and Seawater Intrusion in Coastal Areas of Arid and Semi-Arid Regions. *Water* **2018**, *10*, 143. [CrossRef]
16. Polemio, M. Monitoring and Management of Karstic Coastal Groundwater in a Changing Environment (Southern Italy): A Review of a Regional Experience. *Water* **2016**, *8*, 148. [CrossRef]

*water*  **MDPI**

*Article*

# Atoll Groundwater Movement and Its Response to Climatic and Sea-Level Fluctuations

**Ferdinand K. J. Oberle** [1,*], **Peter W. Swarzenski** [2] **and Curt D. Storlazzi** [1]

1  U.S. Geological Survey, Pacific Coastal and Marine Science Center, Santa Cruz, CA 95060, USA; cstorlazzi@usgs.gov

2  International Atomic Energy Agency, 98000 Monaco, Principality of Monaco; p.swarzenski@iaea.org

*  Correspondence: foberle@usgs.gov; Tel.: +1-831-460-7589

Academic Editors: Maurizio Polemio and Kristine Walraevens
Received: 26 July 2017; Accepted: 22 August 2017; Published: 30 August 2017

**Abstract:** Groundwater resources of low-lying atoll islands are threatened due to short-term and long-term changes in rainfall, wave climate, and sea level. A better understanding of how these forcings affect the limited groundwater resources was explored on Roi-Namur in the Republic of the Marshall Islands. As part of a 16-month study, a rarely recorded island-overwash event occurred and the island's aquifer's response was measured. The findings suggest that small-scale overwash events cause an increase in salinity of the freshwater lens that returns to pre-overwash conditions within one month. The overwash event is addressed in the context of climate-related local sea-level change, which suggests that overwash events and associated degradations in freshwater resources are likely to increase in severity in the future due to projected rises in sea level. Other forcings, such as severe rainfall events, were shown to have caused a sudden freshening of the aquifer, with salinity levels retuning to pre-rainfall levels within three months. Tidal forcing of the freshwater lens was observed in electrical resistivity profiles, high-resolution conductivity, groundwater-level well measurements and through submarine groundwater discharge calculations. Depth-specific geochemical pore water measurements further assessed and confirmed the distinct boundaries between fresh and saline water masses in the aquifer. The identification of the freshwater lens' saline boundaries is essential for a quantitative evaluation of the aquifers freshwater resources and help understand how these resources may be impacted by climate change and anthropogenic activities.

**Keywords:** aquifer; atoll; freshwater lens; sea-level rise; flooding; groundwater; tide; submarine groundwater discharge

## 1. Introduction

In climate change vulnerability assessments, the Marshall Islands as well as the neighboring Kiribati islands, are listed under the "Profound Impacts" category, i.e., the countries "may cease to exist in the event of worst-case scenarios" [1]. A major part of such assessments comes from the limited nature of freshwater resources on low-lying Pacific atoll islands, which is most commonly the critical factor for sustained human habitation. The severity of groundwater dependency was witnessed during a drought in 2016 that caused 16,000 Marshallese, or 30% of the total population to suffer from severe water shortages prompting the Marshallese government to issue a state of emergency [2]. The freshwater resources on low-lying atoll islands typically reside in shallow aquifers, known as freshwater lenses (FWLs), which are naturally recharged only by rainfall and float on top of denser seawater. A brackish transition zone separates saline from fresh water (Figure 1). This hydrogeological setting makes FWLs highly susceptible to vertical mixing that occurs across the entire island and not just at the coastline [3]. In general, the FWL on atoll islands is a function of rainfall, recharge, hydraulic conductivity of the unconsolidated Holocene deposits, and island width, including the reef flat plate and depth to the Thurber discontinuity [4]. The observed FWLs' thicknesses for atoll islands across the Pacific and Indian Ocean are commonly less than 15 m and rarely exceed 20 m [4].

A number of threats to the FWLs have been identified, including: (a) infiltration of anthropogenic contaminants [5]; (b) upconing of saline water due to excessive freshwater pumping [6]; (c) reduction in reef and island size due to coastal erosion, leading in turn to a reduction in the size of the FWL [7]; (d) droughts that hinder successful recharge of the FWLs [5,8–10]; and (e) storm surges that cause large waves to wash over the atolls resulting in saline intrusion [11–13]. Nonetheless, a better understanding of the processes that influence these FWLs—especially in light of expected climate change scenarios on low-lying atoll islands is essential to better assess atoll water resources management challenges in the near future.

The wide range of temporal variability in hydrological processes on low-lying atoll islands complicates the scientific analysis of these processes, while rendering them all the more important. Against the background of expected rising sea levels and more frequent large wave events, hydrogeologic drivers such as tides and altered rainfall patterns must be better understood as they will affect the freshwater resources and consequently lead to a reduction in habitable and cultivatable land. In order to better understand the future changes to the FWLs of low-lying atoll islands, baseline conditions and their temporal variability have to be clearly defined.

The primary goal of this project was to gain a better understanding of the processes that affect the freshwater lens using high-resolution time-series observations of the marine and hydrologic forcings. The effects of multistressors, such as wave-driven overwash events or large rainfall events, represent one of the least monitored and understood topics within atoll hydrology. Only the coupling of hydrological time-series data with oceanographic time-series data will allow a better prediction of future responses of the FWL to the impacts of climate change. Specifically, we present and discuss both geophysical and geochemical data addressing the forcing of the FWL by tides, rainfall, submarine groundwater discharge, large wave events, and high resolution sea-level rise on Roi-Namur Island on Kwajalein Atoll in the Republic of the Marshall Islands.

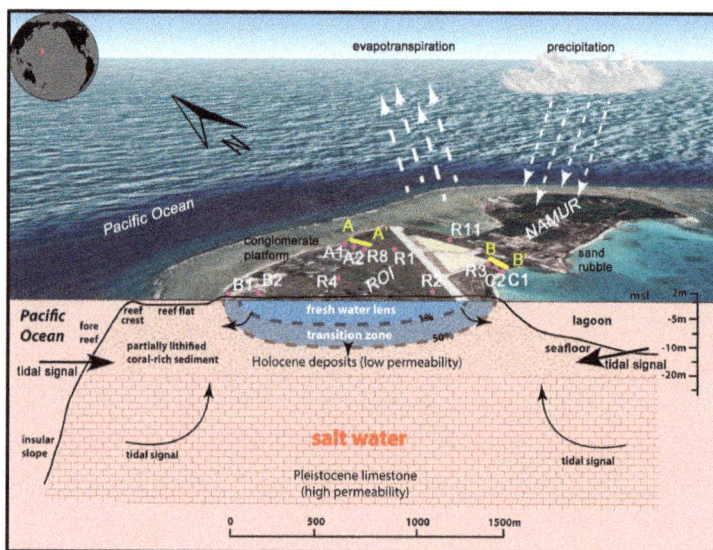

**Figure 1.** Satellite image and conceptual drawing of the shallow aquifer system of Roi-Namur, Kwajalein Atoll, Marshall Islands. Location of shallow groundwater monitoring wells (magenta dots) and time-series electrical resistivity transects (yellow lines) are indicated.

### Study Area

The data presented herein were collected on the island of Roi-Namur (Figure 1) on the northern tip of Kwajalein Atoll in the Republic of the Marshall Islands. Kwajalein is a large (maximum width ~100 km), low-lying (average elevation~2 m) atoll system with a large, deep (>50 m) lagoon and 97 islets that support variably healthy freshwater lenses [14,15]. In 1944, Roi and Namur Islets, located at the northeast tip of the Kwajalein Atoll (lat. 9°23′ N, long. 167°28′ E), were connected by the US Navy with artificial fill to form a single island, now measuring 2.5 km². The reef flat is fully exposed (dry) at low tide, is about 250–350-m wide, and covers an area of about 1 km². Most of the groundwater the water supply system utilizes originates from a horizontal, 1000-m long, skimming well lying just below ground surface, parallel to the runway [13]. This type of well-pumping system limits upconing of the deeper saline water during groundwater withdrawals [16].

Previous research shows the shallow aquifer system at Roi-Namur Island is composed of unconsolidated, reef-derived, calcium-carbonate sand and gravel, with few layers of consolidated rock (coral, sandstone, and conglomerate). The island consists of an approximately 2-m thick disturbed surface layer underlain by three Holocene layers, with a combined thickness of approximately 20 m (Figure 1). This overlays a highly permeable Pleistocene deposit in the order of 900 m thick [14,17]. Aquifer horizontal permeability (k) has previously been calculated to be $1 \times 10^{-11}$–$2 \times 10^{-10}$ m² (hydraulic conductivity [K]: $1 \times 10^{-4}$–$1.6 \times 10^{-3}$ m/s) in the upper Holocene layers and about $3.5 \times 10^{-10}$ m² (K: $3.2 \times 10^{-3}$ m/s) in the lower Pleistocene layer [17]. Roi-Namur's FWL thickness has been shown to vary according to levels of recharge, ranging from 5–12 m thick [13,16]. The groundwater on Roi-Namur is artificially recharged using stored rainwater; this artificial recharge amounts to approximately 3.5% of the natural recharge from rain ($66 \times 10^6$ L/year for the years 2000–2012) and started in 2009 [13]. The available potable freshwater supply has been estimated to $86 \times 10^7$ L for Roi and $16 \times 10^6$ L for Namur [16]. A more general overview on the effects of groundwater pumping on the FWL can be found in Terry et al. [18].

Previous studies [19,20] have demonstrated that global sea level is rising at a rate almost double the Intergovernmental Panel on Climate Change's (IPCC) 2007 report in this area. These high rates of sea-level rise have been tied to strengthened easterly trade winds, which, in turn, appear to be driven by variations in the latent heat content of the earth's warming atmosphere, suggesting that this trend is likely to continue under projected emission scenarios e.g., [21]. Furthermore, the projected sea-level rise will outstrip potential new reef flat accretion, for optimal vertical coral reef flat accretion rates for coral reefs exposed to open-ocean storm waves are up to an order of magnitude smaller (1–4 mm/year per [22,23] than the rates of sea-level rise projected for the years 2000–2100 (8–16 mm/year per [24,25]). For Roi-Namur, this projected scenario results in a net increase in water depth over exposed coral reef flats at the order of 0.4–1.5 m during the 21st century, which will result in larger wave heights [26] and an increase of up to 200% in wave run-up [27], and may ultimately lead to a complete drowning of the islets [28].

## 2. Materials and Methods

### 2.1. Groundwater Levels, Temperature, Specific Conductivity (Salinity) and Water Geochemistry

An assessment of Roi-Namur's shallow freshwater lens was carried out from November 2013 to April 2015. This assessment included surveys of groundwater levels, temperature and specific conductivity (salinity) in a suite of temporary, shallow monitoring wells strategically placed around the island (Figure 1). The wells were constructed of 4-cm-diameter polyvinyl chloride (PVC) pipe with a 60-cm screened section set 15 cm above the bottom to allow groundwater to flow into the well only from the desired depths. Time-series groundwater levels and specific conductivity measurements were performed every 15 min using factory-calibrated Solinst LTC Leveloggers, while time-series groundwater temperature measurements were obtained every 20 min using factory calibrated Onset HOBO temperature loggers. Additionally, depth-specific groundwater samples were collected with an

AMS Piezometer Groundwater Sampling Kit alongside a calibrated YSI 556 multiprobe meter between wells C1 and C2 (Figure 1) using protocols of the USGS National Field Manual [29]. The groundwater samples, that were pumped from a depth of up to 8 m, were analyzed for ammonium ($NH^{+4}$), dissolved silicate (DSi), total dissolved phosphorus (TDP), molybdenum (Mo), barium (Ba), uranium (U), and a suite of hydrological parameters, including pH and salinity. As per methods summarized in Swarzenski et al. [30] nutrients were determined on a Lachat Instruments QuickChem 8000 at Woods Hole Oceanographic Institute (WHOI), while the suite of trace elements was analyzed on a High Resolution Inductively Coupled Plasma Mass Spectrometer at the University of Southern Mississippi. From these measurements, tidal lag and efficiency could also be determined, and is useful for estimating aquifer permeability and storage properties [15].

## 2.2. Electrical Resistivity Tomography (ERT) Surveys

The utility of electrical resistivity to examine the dynamics and scales of the freshwater/saltwater interface in coastal groundwater is well established [31–33]. Time-series multichannel, electrical resistivity tomography (ERT) surveys were conducted along two transects (A–A' and B–B' on Figure 1) during both high and low tides in March 2013. Because the survey cable remained fixed in position on the ground surface during the high tide/low tide and no acquisition parameters were altered during collection, the observed changes in resistivity are only a function of the tidally-modulated pore–fluid exchange. Transects were aligned perpendicularly to the shoreline and located 0.75 m above mean sea level. A SuperSting R8 system (Advanced Geosciences Inc. [AGI], Austin, TX, USA) was used to measure the electrical resistivity of the subsurface along a 56-electrode cable (consistently spaced either 1- or 2-m apart). Each electrode was pinned to the underlying sediment with a 35-cm stainless steel spike. The electrical resistivity measurements were acquired using a dipole-dipole array setting. The relative elevation of each electrode was carefully measured using a Theodelite and the topographic change incorporated into inverse modeling routines (AGI EarthImager).

## 2.3. Submarine Groundwater Discharge (SGD)

Coastal submarine groundwater discharge (SGD) is a highly dynamic and complex hydrogeological phenomenon that involves both terrestrial and marine drivers that define the amount and rate of submarine discharge into the coastal sea, which also incorporates the exchange of water masses through seawater intrusion into the aquifer [34]. Quantification of SGD rates, even in groundwater limited atoll settings, is important to assess groundwater exchange mechanisms and associated constituent fluxes across the island shoreface. The utility of $^{222}$Rn as a water mass tracer is well-proven to study rates of SGD due to its very short half-life (3.8 d) and its multifold enrichment in groundwater relative to surface water [35]. RAD7 radon detection systems were employed to measure Rn in air using a water/air exchanger. This setup allows for a near real-time calculation of the aqueous Rn concentration by measuring the air $^{222}$Rn concentration and knowing the temperature-dependent $^{222}$Rn partitioning coefficient [31,36–40]. A peristaltic pump was used to produce a continuous stream of coastal surface water into the water/air exchanger, while air from the exchanger was continuously pumped into the RAD7 radon monitor. The RAD7 contains a solid-state, planar, Si alpha (PIPS) detector and converts alpha radiation into usable electronic signals that can discriminate various short-lived daughter products (e.g., $^{218}$Po, $^{214}$Po) from $^{222}$Rn [41]. Time-series measurements of nearshore seawater $^{222}$Rn were obtained using a single RAD7 radon monitor setup for 30-min counting intervals. An additional onsite monitoring station was set up at well R3 (Figure1) to establish a $^{222}$Rn groundwater endmember. $^{222}$Rn time-series measurements were taken every 30 minutes for 12 h. The $^{222}$Rn endmember value was established after measurements at peak values had fully leveled ($n = 10$). For the $^{222}$Rn time-series, the surface- and bottom-waters were instrumented with Solinst LTC Leveloggers that continuously measured pressure, conductivity, and temperature of ambient seawater. A simple non-steady state radon mass-balance box model was then employed for calculations of SGD following methods developed by Burnett and Dulaiova [36] and Burnett et al. [35]. In general, this box model accounts for radon sources from (a) total benthic fluxes via submarine groundwater discharge (SGD); (b) diffusion from sediments; and (c) production from dissolved $^{226}$Ra. Radon losses were calculated by including gas evasion, radioactive decay, and mixing with offshore radon-depleted water. In all cases, the excess inventory per time (i.e., differences between source and sink fluxes) was divided by radon concentration in groundwater (i.e., groundwater endmember) to calculate groundwater discharge. Site locations for these surface water time-series deployments were strategically placed based on previous data e.g., [13,16,42], and assumptions on gradients in oceanographic, geologic, and hydrologic controls (Figure 1).

## 3. Results

The ERT profiles clearly identified the FWL in the nearshore environment by recording salinization with depth, as confirmed with deep pore-water sampling (Figure 2). Although the horizontal variability in the ERT profiles may be interpreted as freshwater fingers related to tidally induced convective forcing as has been shown to occur in models [43,44], it is more likely that it is influenced by near-surface variations in conductivity of the soil matrix. However, the overall homogeneity of the aquifer's substrate was found to be relatively uniform as observed by measurements of the hydraulic pressure signal that propagates from the island perimeter inward through the geologic framework of the island.

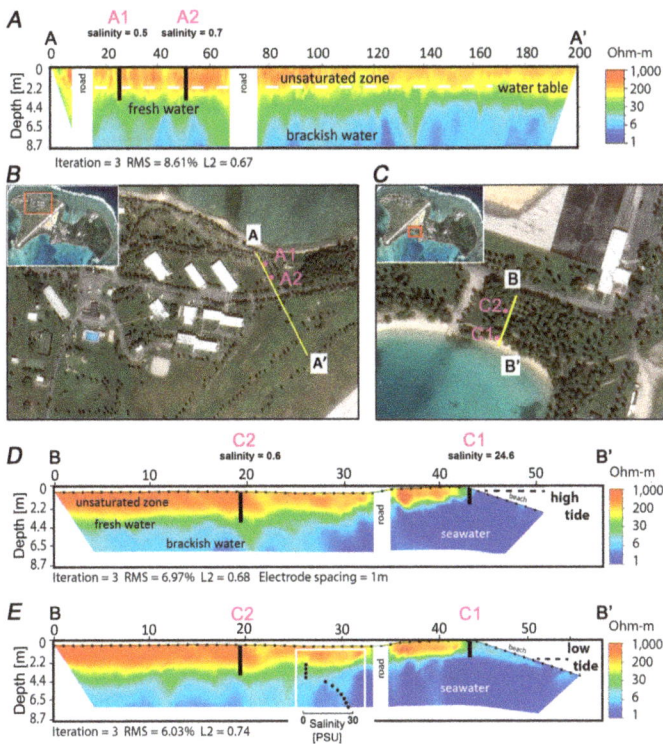

**Figure 2.** Locations and examples of Electrical Resistivity Tomography (ERT) profiles; (**A**) ERT profile A–A' on the ocean side of Roi; (**B**) Map showing location of ERT profile A–A'; (**C**). Map showing location of ERT profile B–B'; (**D**). ERT profile B–B' on the lagoon side of Roi at high tide; (**E**). ERT profile B–B' on the lagoon side of Roi at low tide. The locations of wells A1, A2, C1, and C2 are denoted in magenta. The pore water salinity profile shown in the lower image was obtained using a drive point piezometer, as described in Section 3.1.

Although the ERT profiles clearly identified a distinct FWL floating atop seawater (Figure 2), they were also ground-truthed through geochemical pore water profiles (Figure 3) along the same lagoonal transect (B-B', Figure 2). Most geochemical analyses recorded a sharp transition zone below 4 m depth, indicating a zone of mixing and a differentiation of unique water masses. Salinities sharply increased below 4 m from a freshwater environment (salinity = 1) to a saline environment (salinity = 28). Approximately fourfold similar increases between the water masses were measured in ammonium ($NH^{+4}$), phosphate ($PO_4^{-3}$), molybdenum (Mo), uranium (U), and pH. On the other hand, barium (Ba) and silicate (Si) were highest near the surface and decreased with depth at more gradual rates.

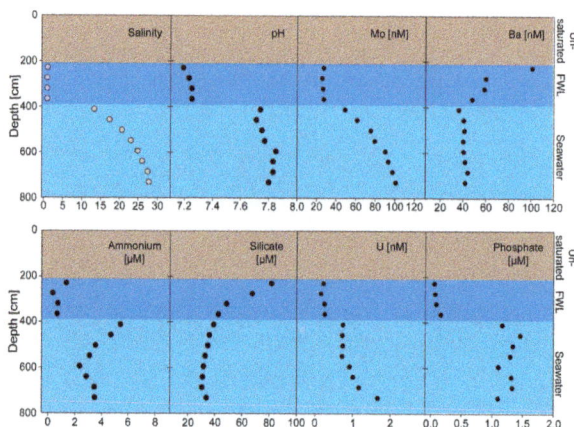

**Figure 3.** Geochemical pore water profiles demonstrating the sharp transition zone between the FWL and the underlying saline groundwater. Site location for these pore water profiles is indicated on Figure 2 between wells C1 and C2.

In order to evaluate the islands permeability and poroelastic storage capacity, a measurement of the attenuation of the tidal signal was captured as tidal efficiency and tidal lag. Tidal efficiency is the amplitude ratio between the magnitudes of the response seen in a well and the corresponding tides. Tidal lag is the time difference or phase lag between the tidal peak and the aquifer response peak. In general, this pressure signal is dampened by frictional losses related to the permeability and the poroelastic storage capacity of the aquifer. Consequently, it is important to assess tidal lag and efficiency in order to address the homogeneity of the aquifer's substrate. Tidal efficiencies from the shallow monitoring wells ranged from 7 to 63%, while the tidal lag ranged from 40 to 170 min for wells measured at 3 m of groundwater depth relative to mean sea level. The measurements of tidal lag and efficiency attained from multiple coastal wells at 3-m depth fit well ($R^2 = 0.9$) with the previously published data by Gingerich [42] (in Peterson [45]) and demonstrate that tidal attenuation is predictably dependent on the distance from the shoreline (Figure 4). Gingerich [42] demonstrated that tidal lag and efficiency were also dependent on the depth of each well, showing that deeper wells had higher efficiencies as the signal traveled through the Pleistocene layer and then upward through the Holocene layer.

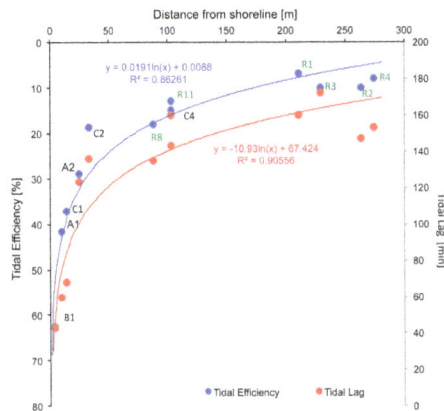

**Figure 4.** Tidal efficiency and tidal lag calculations for groundwater wells. Groundwater well locations shown in Figure 1. Data for groundwater wells R1–R11 (green) were published by Gingerich, 1992 and normalized to 3 m depth using a linear relationship.

## 3.1. Tidal Effect

The dominant driver for the repetitive changes in salinity and groundwater level on Roi-Namur is the oceanic tide (Figure 5A). The measurements show that conductivity can vary up to 10,000 μS/cm within a single tidal cycle (Figure 5B). These changes in conductivity, as a proxy for salinity, are also synchronized with changes in groundwater levels, both temporally and in magnitude. Consequently, spring tides cause the greatest rise of groundwater levels (maximum = 0.75 m) and salinity (maximum conductivity = 10,000 μS/cm), while neap tides are expressed as having a significantly lower groundwater levels (maximum = 0.20 m) and salinities (maximum conductivity = 2000 μS/cm). The ERT profiles also imaged the vertical shift of the FWL with the tidal oscillations. Groundwater wells within the ERT profiles confirmed this by recording freshwater during low tides and in seawater during high tides at their base (Figure 2D,E).

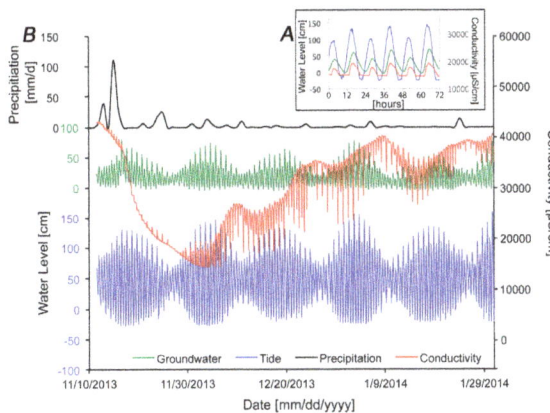

**Figure 5.** Time series plots showing tidal forcing of groundwater levels and salinity (Well C1), as well as the observed response to a large rain event. (**A**) From 11 November 2013 through 29 January 2014. (**B**) From 1 December through 4 December 2013. Precipitation gauge data available from NASA/RTS (2017) [46].

On the same tidally controlled temporal scale, SGD also fluctuated between high and low tide (Figure 6). Although SGD rates were generally quite low (average~1 cm/d), SGD rates varied between 0–3 cm/d. Increased SGD rates were observed shortly after low tide, whereas high tides caused a landward seawater hydraulic head gradient that limited SGD. The ERT profiles confirm this aquifer dynamic (Figure 2D,E), displaying no obvious SGD during high tide but increased brackish discharge during low tide.

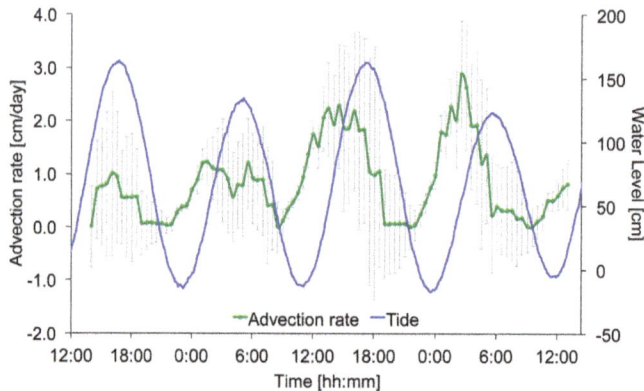

**Figure 6.** Time series showing ocean water levels (tides, in blue) and the computed submarine groundwater discharge (SGD, in green) advection rates. The error bars denote ± 1 standard deviation.

### 3.2. Rainfall Effect

The size of the FWLs of low-lying atolls such as Roi-Namur is largely dependent on the rainfall recharge rate and its temporal variation, such as seasonal and interannual rainfall variability related to the El Niño-Southern Oscillation (ENSO) [8,10] plus groundwater extractions [3,13]. In general however, the rate at which the FWL responds to a large rainfall event depends on the duration and intensity of rainfall. The time-series data captured a large scale rain event (Figure 5B) lasting five days, during which approximately 10% of the annual rainfall (~160–190 cm/year) occurred [13,42]. Such rainfall events are often tied to large low pressure systems and are not uncommon on Roi-Namur [47]. The response to this rain event was recorded in the monitoring wells by a relatively rapid freshening of the aquifer, followed by a gradual increase in conductivity (Figure 5B). The initial decrease in conductivity by 64% (or 25,000 µS/cm) caused by this rain event occurred over a 30-d period, while the return to salinity levels prior to the rain event took approximately 3 mo with little rain (8 cm) falling during this period. The groundwater level observed in the wells was only elevated 25 cm for 24 h directly after the large rainfall. This indicates that increased rates in SGD due to rainfall occurred only for a short time period following the rain event, but were not correlative with the associated decreases or increases in salinity.

### 3.3. Large Wave and Overwash Effect

On 2–3 March 2014, a series of large waves struck Roi-Namur, resulting in wave-driven flooding of the northern portion of the island. The oceanographic forcing that lead to this event has been documented in detail by Quataert et al. [27] and Cheriton et al. [48]. The large wave event, which had almost 7-m high waves with 15-s periods, coincided with a spring high tide and caused ocean water surface elevations, combined with wave run-up, to be 3.7 m above the reef flat, resulting in minor seawater flooding of select low-lying inland areas (Figure 7).

**Figure 7.** Photos of wave-driven flooding and overwash on Roi-Namur Atoll, Republic of the Marshall Islands near the north end of the runway. Photo on the left is looking west and photo on the right is looking east. Red dot on subplot (left photo) indicates the photos location on Roi-Namur.

The event was evidenced in groundwater wells A1, A2, B1, B2, C1, C2, and R3 where it caused an increase in salinity levels e.g., (Figure 8). The overwash event was also accompanied by a sudden and significant rise in groundwater level (20% over the tidally attributed effect) in all wells. The overwash's effect on salinities in wells A1, A2, B1, B2, C1, C2 and R3 was likely dampened by two large rain events that occurred 10 d prior and 1 d (Figure 9) after the overwash event, causing a more gradual signal of increasing salinity than is typical e.g., [9] of saline intrusion. The salinity levels at the nearshore well location A1 closest to the areas flooded with seawater returned to pre-overwash levels within 1 month (Figure 8). Groundwater wells on the lagoonal side of the island returned to pre-overwash salinity levels within 3 d after the overwash event (Figure 9).

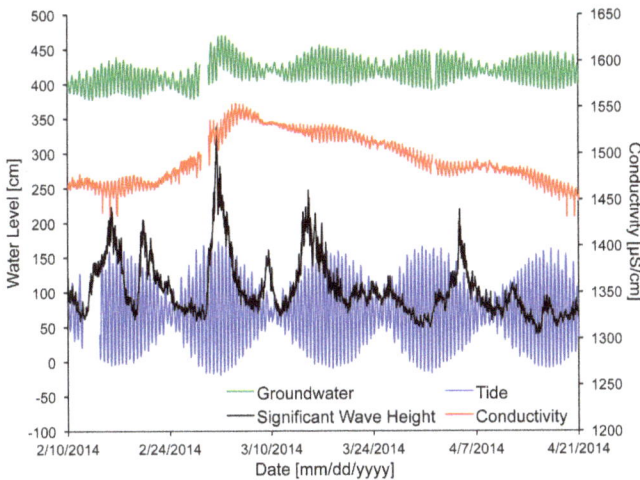

**Figure 8.** Observed variations in conductivity and ground water level in well A1 following an island overwash event driven by increased significant wave height during 2–3 March 2014.

## 3.4. Long-Term Rainfall Effect

With the exception of the two large rain events during November 2013 and March 2014, the collected data support the known rainfall patterns: a relatively dry season from December to April, followed by a wetter season with maximum rainfall occurring between August and October (Figure 9).

During the study period from November 2013 through February 2015, rainfall on Roi-Namur amounted to a total of 274 cm, or 198 cm/year. A substantial portion of the total rainfall occurred in the form of two significant storm events followed by periods of significantly reduced rainfall. The large rain event of November 2013 marked the beginning of what was later to become Typhoon Haiyan near the Philippines, one of the most intense and destructive tropical cyclones on record [49]. The annual fluctuation of salinity in groundwater wells was directly correlative to the annual rainfall patterns, as well as the large storm-induced rainfall events in November 2013 and March 2014 that caused a rapid reduction in conductivity. The three largest rainfall events within the study period occurred on 15 November 2013, 21 February 2014 and 6 September 2014, and were all followed by record minimum salinity levels 30–35 days afterwards. The likelihood of this correlation occurring randomly is less than 1% for the data sets of this study period. In the inland well R3, where oceanographic signals interfere less, a reduction in conductivity of 1000 µS/cm was recorded (Figure 9) after these rainfall events. Although these storm events also caused a sudden increase in groundwater levels, these increases were short-lived (maximum duration of 2 d) and long-term rainfall trends could not be linked to groundwater levels, which steadily decreased over the duration of the study.

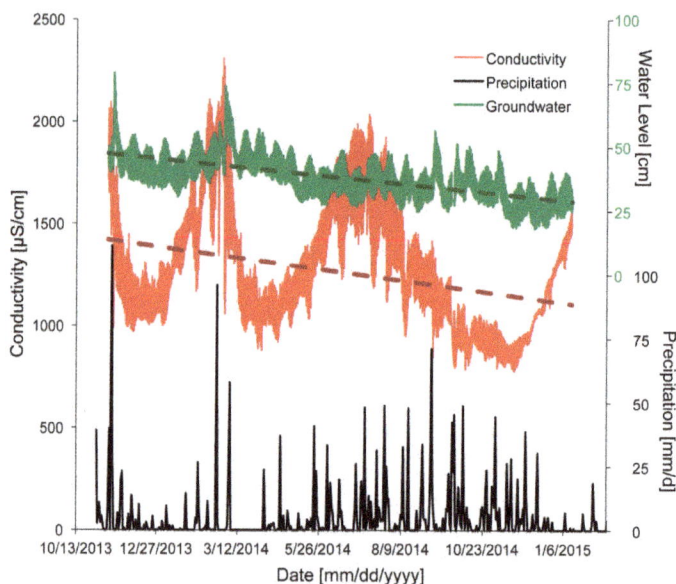

**Figure 9.** Time series plot showing variations in rainfall (black), groundwater level (green), and conductivity of water (red) in well R3 observed between November 2013 and February 2015. Linear fits to the data are denoted by dotted lines.

### 3.5. Sea-Level Change Effect

By comparison to the beginning of the 16-mo monitoring period, the observed salinity and groundwater levels fell by approximately 29% and 68%, respectively (Figure 9). Although a water table is never truly stable, the observed negative trend in groundwater height is closely correlative to the varying levels of decline in mean sea level (average 59%) during this time period (Figure 10A). While mean sea level (MSL) measured by tide gauges since 1948 has, on average, been rising at Roi-Namur by 2.2 mm/year (Figure 10B), previous studies [50,51] have shown that during a significant El Niño-Southern Oscillation (ENSO) event a reduction in local mean sea level can occur. Because the monitoring period of this study fell into a time frame of a strong ENSO event, a decrease in MSL of 13.56 cm total or 11.56 cm/year was observed. Discrepancies between groundwater levels and the fall of MSL were most pronounced during January–March 2014 in Well R11 (Figure 10A). In conjunction with a large decrease in salinity during this time frame (Figure 10A), this discrepancy may indicate the presence of artificial recharge of the FWL as described by Gingerich et al. [13] that would yield higher groundwater levels than from natural sea-level forcing alone. For the study period after March 2014, a cross-correlation analysis of the MSL and the groundwater level data revealed a maximum correlation of 76% at a 5-h shift. However, a definitive answer to this question requires a more detailed analysis of artificial recharge volume, location, and timing.

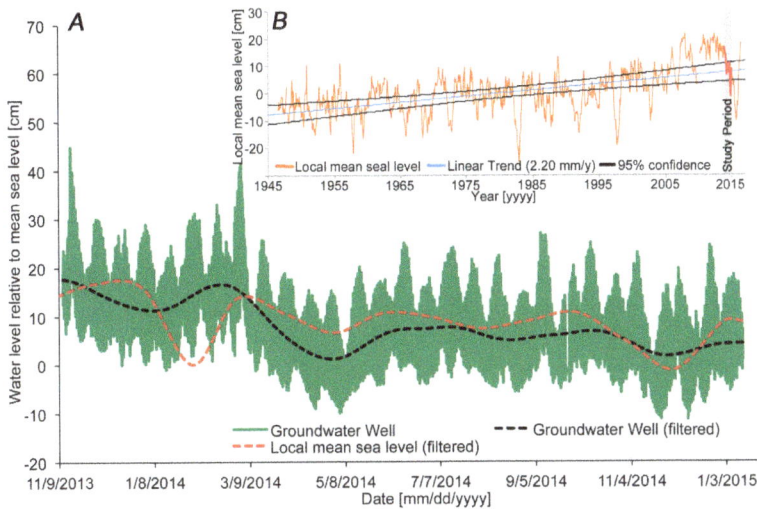

**Figure 10.** Time series of sea-level forcing and well response; (**A**) Comparison of groundwater levels and sea levels during the study at Well R11; (**B**) Local mean sea-level curve from 1945 to 2017. Period pertaining to this study is marked in gray. Sea level data for Kwajalein Atoll from NOAA [52]. A Butterworth filter with a frequency cutoff of 60 d was used to filter data.

## 4. Discussion

High-resolution temporal observations and analyses of hydrological and oceanographic processes are essential for responsible water resources management on atoll islands, and are especially valuable for natural hazard risk reduction. For Roi-Namur, the results of this study indicate that multiple stressors variably affect the freshwater resources. Whereas the size of the FWL is a function of (1) the geological framework and (2) the hydrodynamics of a two-fluid miscible groundwater system e.g., [1], the temporal variability of the atmospheric (rainfall) and oceanographic (tide, wave and sea-level) forcing play a dominant role for both groundwater levels and hydrogeochemistry.

The influence of sea level caused by spring tides forced the greatest range of variability of groundwater levels and conductivity (salinity) levels, whereas neap tides drive significantly lower variability in groundwater level and conductivity (salinity) levels. The hourly changes in groundwater levels and conductivity (salinity) levels caused by the semidiurnal tidal pressure wave were also the most apparent groundwater signal in all collected datasets e.g., (Figures 5, 6 and 8). This dynamic is expected to be a common response on low-lying carbonate atoll islands and confirms previous studies on tide-induced fluctuations of salinity and groundwater level in unconfined aquifers [53].

The tidal measurements represent a vertically and horizontally averaged response of the aquifer to a pressure wave passing through an unknown section of the aquifer to reach the well. Gingerich [42] previously showed that tidal efficiency is a function of the dual aquifer effect of the more permeable Pleistocene limestone topped by lesser permeable Holocene deposits (Figure 1). Although those results focused on the propagation of the tide with increasing aquifer depth, the results presented here, which were normalized to a common depth of 3 m, show that the tidal lag and efficiency are also strongly dependent on the distance to shoreline (Figure 4). Although this may appear obvious, the significant logarithmic fit ($R^2$ = 90) of the tidal lag and efficiency to shoreline distance implies a relatively high homogeneity of the poroelastic aquifer's composition.

Detailed analyses of sea level forces are specifically important in environments with expected low SGD because tidal oscillation may considerably increase the average recirculated seawater component of coastal groundwater [54]. The relatively low SGD rates for Roi-Namur (mean = 1 ± 3 cm/d) fit with previously published SGD rates from low-lying atoll islands with similar atmospheric and oceanographic forcing [55]. For the southern Pacific Rarotonga Atoll, which is considerably higher but experiences similar levels of rainfall (190 cm) and oceanographic forcing, recent SGD calculations suggest similar flow rates of 0.2–1.8 cm/d [56]. The variable SGD flow rates between high tide and low tide are known as tidal pumping, and are the main driving force of pore water advection. SGD-induced tidal pumping has previously been described as the "breathing" of offshore coral islands, where seawater is inhaled at high tide and nutrient rich groundwater is exhaled at low tide, leading to sustained productivity within coral reefs [57].

Furthermore, the analysis of tidally driven SGD is also important for a more complete understanding of aquifer biogeochemistry and reef health. SGD has been shown to play a major ecological factor connecting the reef to the subsurface environment, which, may in turn lead to connections to land e.g., [41,58,59]. The geochemical loading from SGD has been described as comparable to or exceeding those of surface runoff inputs, due to higher dissolved solids concentrations in groundwater and larger accumulative discharge zones [41,60]. Contrary to these findings, nutrient levels such as phosphate or ammonium from freshwater SGD on Roi-Namur are roughly four times lower than the seawater underlying the island (Figure 3). The increasing levels of ammonium with depth is likely caused by dissimilatory nitrate reduction that occurs in the more anoxic conditions in the deeper parts of the aquifer [61]. The phosphate enrichment in the deeper saline waters have previously been associated with carbonate dissolution. Burt [62] showed that saltwater penetrating a carbonate aquifer is enriched with ammonia and phosphate, relative to fresh groundwater end-members. He attributed these enrichments to calcite dissolution processes of the carbonate aquifer structure driven by $CO_2$ infiltration. On Roi-Namur, Hejazian et al. [16] have shown that the carbonate dissolution leads to increased porosity of the aquifer and a downward transport of undersaturated waters. Similar sharp increases with depth in Mo, U, and pH are likely reflective of an oxygenated freshwater environment (<4 m depth) and a more anoxic saline environment (>4 m depth). While the geochemical data clearly suggest a separation of the FWL from the underlying seawater at 4 m depth, some minor freshwater mixing that decreases with depth is indicated within the more anoxic saline environment (>4 m depth) by salinity, Mo or Si levels. The high surficial levels in Ba and Si imply a surficial anthropogenic source, such as the building materials used in the construction of the runway on Roi-Namur. This would also explain their gradual reduction with depth. In general, this indicates that freshwater SGD is not likely to cause a significant nutrient input to nearshore waters, whereas the tidal flushing of deeper saline groundwater may contribute to nutrients to the reef.

Contrary to the rhythmic, predictable hydrological processes driven by tidal oscillations, storm events and their associated hydrological drivers, such as rainfall and wave-driven flooding, occur less predictably. The high temporal variability of rainfall on Roi-Namur observed during this study confirmed previous analysis of a relatively dry season from December to April, followed by a wetter season with maximum rainfall occurring between August and October. This study shows that heavy rainfall events, such as those caused by storm systems, can significantly and suddenly alter the conductivity (salinity) levels in groundwater wells (Figure 5), while periods of drought or reduced rainfall cause a gradual increase in groundwater wells' salinities (Figure 9). On Roi-Namur and elsewhere in Marshall Islands, this can become particularly important, as rainfall can decrease by 80% during El Niño droughts [63], causing more than half of the Marshall Islands to be classified as "highly vulnerable" to freshwater stress [10]. Model calculations have also shown that during drought, groundwater on small islands (<300 m width) can be completely depleted [10]. In addition, the most recent predictions by the Department of Defense's Strategic Environmental Research and Development Program [64] forecast significant decreases in rainfall. The latter prediction is also supported by measured rainfall data since 1950 that shows a decreasing trend in annual and seasonal rainfall [65]. Nonetheless, there still exists uncertainty around the rainfall projections and not all models show consistent results [47].

Although Roi-Námur lies within a climatic zone where rainfall and storms are likely to occur during specific times of the year, their exact temporal occurrence, specifically in relation to regional sea level and tidal oscillations, is harder to predict. Consequently, concurrent high wave events and spring tides leading to island overwash and seawater flooding occur with little warning [48]. Although the overwash event recorded in this study caused conductivity (salinity) increases that recovered after approximately one month (Figure 8) in affected groundwater wells, more severe overwash events have the potential to be catastrophic, leading to FWL recovery times of 22–26 months [11–13]. For example, in 2008, 2009, and 2011, storm-driven large wave events that coincided with high tides negatively affected freshwater drinking supplies, destroyed vital crops, demolished infrastructure, and killed hundreds of thousands of Federally-protected animal species on Pacific atolls [12,66], highlighting the exceptional vulnerability of these low-lying island communities. We suggest that future work should focus on further developing wave run-up forecasting tools for Roi-Namur and other Pacific Islands [67] and integrating quantitative saline intrusion predictions, thereby not only protecting buildings and general infrastructure, but also the islands' vital freshwater resources.

The recorded conductivity data indicate that the overarching control on salinity levels detected in groundwater wells is not driven by rainfall alone, but a combination of changes in rainfall and sea level. Although tidal pressure causes oscillations in wells' conductivity (salinity) profiles, the longer-term control on salinity levels in groundwater wells can be attributed to a local change in MSL. This can be observed in the slope of the recorded conductivity data that is congruent with the average decline in MSL over the course of this study (Figure 9). The observed dependence of groundwater conductivity levels on MSL indicates that salinity levels in groundwater wells should also increase in the future in accordance with the predicted future rise in MSL.

Mean sea level is projected to continue to rise over the course of the 21st century and there is very high confidence in the direction of change [68]. The most current models simulate a rise in mean sea level between 7 and 19 cm by 2030 using the full range of emission scenarios. Increases in mean sea level of 41–92 cm by 2090 under the Representative Concentration Pathways emission scenario, 8.5 can be expected [69]. The predicted rise in eustatic sea level will also increase the size of large waves over coral reefs and the resulting wave-driven run-up, leading to more frequent and intense overwash events in the future [26]. Although the documented overwash event in this study occurred over a relatively short timeframe (3 h) when the spring tide and the high wave event overlapped, the predicted future sea-level rise will allow overwash events to occur during a much longer portion of the tidal cycle. This also means that in the future, as sea-level rise continues, smaller wave events, which also occur more frequently, will be able to cause island flooding and negatively impact the island's freshwater resources.

## 5. Conclusions

FWLs on small low-lying atoll islands such as Roi-Namur are under constant and ever-increasing threat of salinization. Yet, it is these groundwater resources that are critical for the survival of the local population. The potential threats to the salinity levels of the FWLs are abundant and occur in the form of high tides—especially king tides, lack of rainfall, large wave events, and sea-level rise. Other anthropogenic threats not considered in this study include contamination, over-extraction, and island modification.

This study provides measured responses of conductivity (salinity) in groundwater wells to severe rainfall events, as well as seasonal rainfall patterns. The study provides calculations on tidal lag and efficiency in wells normalized to 3-m depth that demonstrate tidal attenuation is predictably dependent on the distance from the shoreline. An island overwash event was documented and the nearshore aquifer responses identified. ERT profiles characterized the freshwater outflow patterns and the FWL's saline boundaries. Depth-specific geochemical pore water measurements characterized and confirm the distinct boundaries between fresh and saline water masses in the aquifer. SGD was calculated and put in the context of tidal oscillations to show its varying degrees of output. Conductivity (salinity) and groundwater levels were shown to respond to measured mean local sea-level change.

The results show that the impact of the individual threats on the FWLs is strongly dependent on their relative timing. As expected, spring tides were observed to cause the greatest rise of groundwater levels and conductivity (salinity), whereas neap tides are expressed as having a significantly lower groundwater levels and salinities. Tidal oscillations could be identified in both groundwater levels and conductivity (salinity) levels. The tidal signal was also expressed in SGD, which ceased to exist during high tides, effectively allowing saline waters to intrude the outer limits of the FWL, while low tides allowed freshwater to be expelled into the lagoon. Tidal lag and efficiency were reflective of a highly permeable substrate that was homogenous throughout the of the island's nearshore surficial aquifer strata. Large rain events were shown to have a relatively sudden freshening effect in groundwater wells (within 3 d), while returns to initial conductivity (salinity) levels took up to an order of magnitude longer (30 d). The observed overwash event caused short-lived increases in groundwater levels and an increase in conductivity (salinity) levels that recovered after 1 mo; this event was minor by comparison to previous events that had recovery times of up to 26 mo. Groundwater levels and salinity levels were observed to follow the general falling trend in MSL caused by an El Niño-Southern Oscillation event during this study. Together, these observations demonstrate the atoll aquifer's response to a wide range of atmospheric and oceanographic forcing over a range of temporal scales, and provides insight into the vulnerability of the FWL that will likely heighten with the predicted changes in rainfall and sea level due to global climate change.

**Acknowledgments:** Funding for this research was provided by the U.S. Department of Defense's Strategic Environmental Research and Development Program (SERDP) under Project RC-2334 and the U.S. Geological Survey's Coastal and Marine Geology Program. The IAEA is grateful for the support provided to its Environment Laboratories by the Government of the Principality of Monaco. We thank S.B. Gingerich for his foresight in developing a groundwater monitoring network on Roi-Namur. We also thank C. Johnson and K.O. Odigie for their help with instrument deployment and data collection. We also thank the Captain and crew of the D/V Patriot, D. Miller and C. Nakasone (Kwajalein Range Services), and the U.S. Army Garrison-Kwajalein Atoll (USAG-KA) for their support of this project. Any use of trade, firm, or product names is for descriptive purposes only and does not imply endorsement by the U.S. Government.

**Author Contributions:** F.K.J.O., P.W.S. and C.D.S. conceived and designed the experiments; F.K.J.O. and P.W.S. and C.D.S. performed the experiments; F.K.J.O. and P.W.S. analyzed the data; F.K.J.O., P.W.S. and C.D.S. wrote the paper. Authorship must be limited to those who have contributed substantially to the work reported.

**Conflicts of Interest:** The authors declare no conflicts of interest. The founding sponsors had no role in the design of the study; in the collection, analyses, or interpretation of data; in the writing of the manuscript, and in the decision to publish the results.

## References

1. Pernetta, J.C. Impacts of climate change and sea-level rise on small island states. *Glob. Environ. Chang.* **1992**, *2*, 19–31. [CrossRef]
2. Government of the Marshall Islands. Immediate Drought Response Plan For the Republic of the Marshall Islands Complementing the Declaration on State of Emergency. Available online: http://reliefweb.int/report/marshall-islands/immediate-drought-response-plan-republic-marshall-islands (accessed on 26 August 2017).
3. Underwood, M.R.; Peterson, F.L.; Voss, C.I. Groundwater lens dynamics of Atoll Islands. *Water Resour. Res.* **1992**, *28*, 2889–2902. [CrossRef]

4.  Bailey, R.T.; Jenson, J.W.; Olsen, A.E. Estimating the Ground Water Resources of Atoll Islands. *Water* **2010**, *2*, 1–27. [CrossRef]
5.  White, I.; Falkland, T.; Perez, P.; Dray, A.; Metutera, T.; Metai, E.; Overmars, M. Challenges in freshwater management in low coral atolls. *J. Clean. Prod.* **2007**, *15*, 1522–1528. [CrossRef]
6.  Falkland, A.; Custodio, E.; Diaz Arenas, E.; Simler, E. *Hydrology and Water Resources of Small Islands: A Practical Guide*; UNESCO: Paris, France, 1991; Volume 49, ISBN 9231027530.
7.  Terry, J.P.; Thaman, R.R. Physical geography of Majuro and the Marshall Islands. In *The Marshall Islands: Environment, History and Society in the Atolls*; Faculty of Islands and Oceans, the University of the South Pacific: Suva, Fiji, 2008; pp. 1–22.
8.  Van der Velde, M.; Javaux, M.; Vanclooster, M.; Clothier, B.E. El Niño-Southern Oscillation determines the salinity of the freshwater lens under a coral atoll in the Pacific Ocean. *Geophys. Res. Lett.* **2006**, *33*, L21403. [CrossRef]
9.  Parry, M.L.; Canziani, O.F.; Palutikof, J.P.; van der Linden, P.J.; Hanson, C.E. Small islands. Climate Change. Climate Change 2007: Impacts, Adaptation and Vulnerability. Contribution of Working Group II to the Fourth Assessment Report of the Intergovernmental Panel on Climate Change. In *Climate Change 2007: Impacts, Adaptation and Vulnerability*; Cambridge University Press: Cambridge, UK, 2007; pp. 671–687.
10. Barkey, B.; Bailey, R. Estimating the Impact of Drought on Groundwater Resources of the Marshall Islands. *Water* **2017**, *9*, 41. [CrossRef]
11. Terry, J.P.; Falkland, A.C. Responses of atoll freshwater lenses to storm-surge overwash in the Northern Cook Islands. *Hydrogeol. J.* **2010**, *18*, 749–759. [CrossRef]
12. Fletcher, B.C.H.; Richmond, B.M. *Report of Findings—Climate Change in the Federated States of Micronesia: Food and Water Security, Climate Risk Management, and Adaptive Strategies*; University of Hawai'i Sea Grant College Program: Honolulu, HI, USA, 2010; pp. 1–32.
13. Gingerich, S.B.; Voss, C.I.; Johnson, A.G. Seawater-flooding events and impact on freshwater lenses of low-lying islands: Controlling factors, basic management and mitigation. *J. Hydrol.* **2017**, *551*, 676–688. [CrossRef]
14. Gingerich, S.B. Groundwater resources and contamination at RoiNamur Island, Kwajalein Atoll, Republic of the Marshall Islands, 1990–91. *Water-Resour. Investig. Rep.* **1996**, *95*, 1–10.
15. Hunt, C.D. Ground-Water Resources and Contamination at Kwajalein Island, Republic of the Marshall Islands, 1990–91. *Water Resour. Investig. Rep.* **1996**, *94*, 1–10.
16. Hejazian, M.; Gurdak, J.J.; Swarzenski, P.; Odigie, K.O.; Storlazzi, C.D. Land-use change and managed aquifer recharge effects on the hydrogeochemistry of two contrasting atoll island aquifers, Roi-Namur Island, Republic of the Marshall Islands. *Appl. Geochem.* **2017**, *80*, 58–71. [CrossRef]
17. Peterson, F.L.; Gingerich, S.B. Modeling atoll groundwater systems. In *Groundwater Models for Resources Analysis and Management*; El-Kadi, A.I., Ed.; CRC Press: Boca Raton, FL, USA, 1995; pp. 275–292.
18. Terry, J.P.; Chui, T.F.M.; Falkland, A. Atoll Groundwater Resources at Risk: Combining Field Observations and Model Simulations of Saline Intrusion Following Storm-Generated Sea Flooding. In *Groundwater in the Coastal Zones of Asia-Pacific*; Wetzelhuetter, C., Ed.; Springer: Dordrecht, The Netherlands, 2013; pp. 247–270. ISBN 978-94-007-5648-9.
19. Merrifield, M.A.; Merrifield, S.T.; Mitchum, G.T. An anomalous recent acceleration of global sea level rise. *J. Clim.* **2009**, *22*, 5772–5781. [CrossRef]
20. Vermeer, M.; Rahmstorf, S. Global sea level linked to global temperature. *Proc. Natl. Acad. Sci. USA* **2009**, *106*, 21527–21532. [CrossRef] [PubMed]
21. Merrifield, M.A. A shift in western tropical Pacific sea level trends during the 1990s. *J. Clim.* **2011**, *24*, 4126–4138. [CrossRef]
22. Buddemeier, R.W.; Smith, S.V. Coral reef growth in an era of rapidly rising sea level: Predictions and suggestions for long-term research. *Coral Reefs* **1988**, *7*, 51–56. [CrossRef]
23. Montaggioni, L.F. History of Indo-Pacific coral reef systems since the last glaciation: Development patterns and controlling factors. *Earth-Sci. Rev.* **2005**, *71*, 1–75. [CrossRef]
24. Grinsted, A.; Moore, J.C.; Jevrejeva, S. Reconstructing sea level from paleo and projected temperatures 200 to 2100 ad. *Clim. Dyn.* **2010**, *34*, 461–472. [CrossRef]
25. Nicholls, R.J.; Cazenave, A. Sea-Level Rise and Its Impact on Coastal Zones. *Science* **2010**, *328*, 1517–1520. [CrossRef] [PubMed]

26. Storlazzi, C.D.; Shope, J.B.; Erikson, L.H.; Hegermiller, C.A.; Barnard, P.L. Future wave and wind projections for U.S. and U.S.-affiliated pacific islands. *U.S. Geol. Surv. Open-File Rep.* **2015**, *1001*, 1–426.

27. Quataert, E.; Storlazzi, C.; Van Rooijen, A.; Cheriton, O.; Van Dongeren, A. The influence of coral reefs and climate change on wave-driven flooding of tropical coastlines. *Geophys. Res. Lett.* **2015**, *42*, 6407–6415. [CrossRef]

28. Albert, S.; Leon, J.X.; Grinham, A.R.; Church, J.A.; Gibbes, B.R.; Woodroffe, C.D. Interactions between sea-level rise and wave exposure on reef island dynamics in the Solomon Islands. *Environ. Res. Lett.* **2016**, *11*, 54011. [CrossRef]

29. U.S. Geological Survey. *National Field Manual for the Collection of Water-Quality Data*; U.S. Geological Survey: Reston, VA, USA, 1998; Book 9, Chapters A1–A9.

30. Swarzenski, P.W.; Simonds, F.W.; Paulson, A.J.; Kruse, S.; Reich, C. Geochemical and Geophysical Examination of Submarine Groundwater Discharge and Associated Nutrient Loading Estimates into Lynch Cove, Hood Canal, WA. *Environ. Sci. Technol.* **2007**, *41*, 7022–7029. [CrossRef] [PubMed]

31. Swarzenski, P.W.; Burnett, W.C.; Greenwood, W.J.; Herut, B.; Peterson, R.; Dimova, N.; Shalem, Y.; Yechieli, Y.; Weinstein, Y. Combined time-series resistivity and geochemical tracer techniques to examine submarine groundwater discharge at Dor Beach, Israel. *Geophys. Res. Lett.* **2006**, *33*, L24405. [CrossRef]

32. Swarzenski, P.W.; Kruse, S.; Reich, C.; Swarzenski, W.V. Multi-channel resistivity investigations of the freshwater-saltwater interface: A new tool to study an old problem. In Proceedings of the International Symposium: A New Focus on Groundwater—Seawater Interactions, Perugia, Italy, 2–13 July 2007; pp. 1–7.

33. Manheim, F.T.; Krantz, D.E.; Bratton, J.F. Studying Ground Water Under Delmarva Coastal Bays Using Electrical Resistivity. *Ground Water* **2004**, *42*, 1052–1068. [CrossRef]

34. Zektser, I.S.; Dzyuba, A.V. Submarine discharge into the Barents and White Seas. *Environ. Earth Sci.* **2014**, *71*, 723–729. [CrossRef]

35. Swarzenski, P.W. U/Th Series Radionuclides as Coastal Groundwater Tracers. *Chem. Rev.* **2007**, *107*, 663–674. [CrossRef] [PubMed]

36. Burnett, W.C.; Dulaiova, H. Estimating the dynamics of groundwater input into the coastal zone via continuous radon-222 measurements. *J. Environ. Radioact.* **2003**, *69*, 21–35. [CrossRef]

37. Burnett, W.C.; Bokuniewicz, H.; Huettel, M.; Moore, W.; Taniguchi, M. Groundwater and pore water inputs to the coastal zone. *Biogeochemistry* **2003**, *66*, 3–33. [CrossRef]

38. Burnett, W.C.; Aggarwal, P.K.; Aureli, A.; Bokuniewicz, H.; Cable, J.E.; Charette, M.A.; Kontar, E.; Krupa, S.; Kulkarni, K.M.; Loveless, A.; et al. Quantifying submarine groundwater discharge in the coastal zone via multiple methods. *Sci. Total Environ.* **2006**, *367*, 498–543. [CrossRef] [PubMed]

39. Dulaiova, H.; Burnett, W.C.; Chanton, J.P.; Moore, W.S.; Bokuniewicz, H.J.; Charette, M.A.; Sholkovitz, E. Assessment of groundwater discharges into West Neck Bay, New York, via natural tracers. *Cont. Shelf Res.* **2006**, *26*, 1971–1983. [CrossRef]

40. Schubert, M.; Paschke, A.; Lieberman, E.; Burnett, W.C. Air–Water Partitioning of $^{222}$Rn and its Dependence on Water Temperature and Salinity. *Environ. Sci. Technol.* **2012**, *46*, 3905–3911. [CrossRef] [PubMed]

41. Swarzenski, P.W.; Dulaiova, H.; Dailer, M.L.; Glenn, C.R.; Smith, C.G.; Storlazzi, C.D. *A Geochemical and Geophysical Assessment of Coastal Groundwater Discharge at Select Sites in Maui and O'ahu, Hawai'i*; Springer: Amsterdam, The Netherlands, 2015; Volume 7, pp. 27–46. ISBN 9789400756472.

42. Gingerich, S.B. Numerical Simulation of the Freshwater Lens on Roi-Namur Island, Kwajalein Atoll, Repubic of the Marshall Islands. Master's Thesis, University of Hawaii, Honolulu, HI, USA, 1992.

43. Greskowiak, J. Tide-induced salt-fingering flow during submarine groundwater discharge. *Geophys. Res. Lett.* **2014**, *41*, 6413–6419. [CrossRef]

44. Kooi, H.; Groen, J.; Leijnse, A. Modes of seawater intrusion during transgressions. *Water Resour. Res.* **2000**, *36*, 3581–3589. [CrossRef]

45. Peterson, F.L. Hydrogeology of the Marshall Islands. In *Geology and Hydrogeology of Carbonate Islands. Developments in Sedimentology*; Elsevier: Amsterdam, The Netherlands, 1997; Volume 54, pp. 611–636. ISBN 9781627034470.

46. NASA and RTS Precipitation Measurment Mission Ground Validation. Available online: https://trmm-fc.gsfc.nasa.gov/trmm_gv/data/data.html (accessed on 26 August 2017).

47. Australian Bureau of Meteorology (ABM); Commonwealth Scientific and Industrial Research Organisation (CSIRO). *Climate Variability, Extremes and Change in the Western Tropical Pacific: New Science and Updated Country Reports 2014*; Centre for Australian Weather and Climate Research: Melbourne, Australia, 2014.

48. Cheriton, O.M.; Storlazzi, C.D.; Rosenberger, K.J. Observations of wave transformation over a fringing coral reef and the importance of low-frequency waves and offshore water levels to runup, overwash, and coastal flooding. *J. Geophys. Res. Oceans* **2016**, *121*, 3121–3140. [CrossRef]

49. Dolan, C.J.; Lyon, A.J. Calculation of Goodwill: Humanitarianism, Strategic Interests, and the U.S. Response to Typhoon Yolanda. *Glob. Secur. Intell. Stud.* **2016**, *2*. [CrossRef]

50. Chowdhury, M.R.; Chu, P.-S.; Schroeder, T. ENSO and seasonal sea-level variability—A diagnostic discussion for the U.S.-Affiliated Pacific Islands. *Theor. Appl. Climatol.* **2007**, *88*, 213–224. [CrossRef]

51. Becker, M.; Meyssignac, B.; Letetrel, C.; Llovel, W.; Cazenave, A.; Delcroix, T. Sea level variations at tropical Pacific islands since 1950. *Glob. Planet. Chang.* **2012**, *80–81*, 85–98. [CrossRef]

52. NOAA (National Oceanic and Atmospheric Administration). Mean Sea Level Trend Kwajalein, Pacific Ocean. Available online: https://tidesandcurrents.noaa.gov/sltrends/sltrends_station.shtml?stnid=1820000 (accessed on 26 August 2017).

53. Levanon, E.; Yechieli, Y.; Gvirtzman, H.; Shalev, E. Tide-induced fluctuations of salinity and groundwater level in unconfined aquifers—Field measurements and numerical model. *J. Hydrol.* **2016**, *551*, 665–675. [CrossRef]

54. Prieto, C.; Destouni, G. Quantifying hydrological and tidal influences on groundwater discharges into coastal waters. *Water Resour. Res.* **2005**, *41*, 1–12. [CrossRef]

55. Moosdorf, N.; Stieglitz, T.; Waska, H.; Dürr, H.H.; Hartmann, J. Submarine groundwater discharge from tropical islands: A review. *Grundwasser* **2014**, *20*, 53–67. [CrossRef]

56. Tait, D.R.; Santos, I.R.; Erler, D.V.; Befus, K.M.; Cardenas, M.B.; Eyre, B.D. Estimating submarine groundwater discharge in a South Pacific coral reef lagoon using different radioisotope and geophysical approaches. *Mar. Chem.* **2013**, *156*, 49–60. [CrossRef]

57. Santos, I.R.; Erler, D.; Tait, D.; Eyre, B.D. Breathing of a coral cay: Tracing tidally driven seawater recirculation in permeable coral reef sediments. *J. Geophys. Res. Oceans* **2010**, *115*, 1–10. [CrossRef]

58. Knee, K.L.; Crook, E.D.; Hench, J.L.; Leichter, J.J.; Paytan, A. Assessment of Submarine Groundwater Discharge (SGD) as a Source of Dissolved Radium and Nutrients to Moorea (French Polynesia) Coastal Waters. *Estuaries Coasts* **2016**, *39*, 1651–1668. [CrossRef]

59. Cardenas, M.B.; Zamora, P.B.; Siringan, F.P.; Lapus, M.R.; Rodolfo, R.S.; Jacinto, G.S.; San Diego-McGlone, M.L.; Villanoy, C.L.; Cabrera, O.; Senal, M.I. Linking regional sources and pathways for submarine groundwater discharge at a reef by electrical resistivity tomography, $^{222}$Rn, and salinity measurements. *Geophys. Res. Lett.* **2010**, *37*. [CrossRef]

60. Taniguchi, M.; Burnett, W.C.; Cable, J.E.; Turner, J.V. Investigation of submarine groundwater discharge. *Hydrol. Process.* **2002**, *16*, 2115–2129. [CrossRef]

61. Korom, S.F. Natural denitrification in the saturated zone: A review. *Water Resour. Res.* **1992**, *28*, 1657–1668. [CrossRef]

62. Burt, R.A. Ground-water chemical evolution and diagenetic processes in the upper Floridan Aquifer, southern South Carolina and northeastern Georgia. *USGS Water Supply Pap.* **1993**, *2392*, 1–76.

63. Presley, T.K. *Majuro Water and Sewer Company, Majuro Atoll, Republic of the Marshall Islands Effects of the 1998 Drought on the Freshwater Lens in the Laura Area, Majuro Atoll, Republic of the Marshall Islands*; Scientific Investigations Report 2005-5098; Geological Survey (U.S.): Reston, VA, USA, 2005; 40p.

64. Storlazzi, C.D. SERDP Project RC-2334: The Impact of Sea-Level Rise and Climate Change on Department of Defense Installations on Atolls in the Pacific Ocean. Available online: https://www.serdp-estcp.org/Program-Areas/Resource-Conservation-and-Resiliency/Infrastructure-Resiliency/Vulnerability-and-Impact-Assessment/RC-2334/RC-2334 (accessed on 26 August 2017).

65. Australian Bureau of Meteorology (ABM); Commonwealth Scientific and Industrial Research Organisation (CSIRO). *Current and Future Climate of the Marshall Islands*; CSIRO, Australian Bureau of Meteorology: Melbourne, Australia, 2015; pp. 1–11.

66. Reynolds, M.H.; Courtot, K.N.; Berkowitz, P.; Storlazzi, C.D.; Moore, J.; Flint, E. Will the effects of sea-level rise create ecological traps for Pacific island seabirds? *PLoS ONE* **2015**, *10*, e0136773. [CrossRef] [PubMed]

67.    PacIOOS Wave Run-Up Forecast: Kwajalein, RMI. Available online: http://www.pacioos.hawaii.edu/shoreline/runup-kwajalein/ (accessed on 26 August 2017).

68.    Van Vuuren, D.P.; Edmonds, J.; Kainuma, M.; Riahi, K.; Thomson, A.; Hibbard, K.; Hurtt, G.C.; Kram, T.; Krey, V.; Lamarque, J.F.; et al. The representative concentration pathways: An overview. *Clim. Chang.* **2011**, *109*, 5–31. [CrossRef]

69.    Science, P.C.C.; Program, A.P. *Pacific-Australia Climate Change Science and Adaptation Planning Program Climate Variability, Extremes and Change in the Western Tropical Pacific: New Science and Updated Country Reports*; Australian Bureau of Meteorology and Commonwealth Scientific and Industrial Research Organisation (CSIRO): Canberra, Australia, 2014; ISBN 9781486302888.

*water*

MDPI

*Article*

# Delineation of Salt Water Intrusion through Use of Electromagnetic-Induction Logging: A Case Study in Southern Manhattan Island, New York

Frederick Stumm * and Michael D. Como

U.S. Geological Survey, New York Water Science Center, 2045 Route 112, Coram, NY 11727, USA; mcomo@usgs.gov
* Correspondence: fstumm@usgs.gov

Academic Editors: Maurizio Polemio and Kristine Walraevens
Received: 12 May 2017; Accepted: 19 August 2017; Published: 23 August 2017

**Abstract:** Groundwater with chloride concentrations up to 15,000 mg/L has intruded the freshwater aquifer underlying southern Manhattan Island, New York. Historical (1940–1950) chloride concentration data of glacial aquifer wells in the study area indicate the presence of four wedges of saltwater intrusion that may have been caused by industrial pumpage. The limited recharge capability of the aquifer, due to impervious surfaces and the 22.7 million liters per day (mld) of reported industrial pumpage early in the 20th Century was probably the cause for the saltwater intrusion and the persistence of the historical saltwater intrusion wedges over time. Recent drilling of wells provided new information on the hydrogeology and extent of saltwater intrusion of the glacial aquifer overlying bedrock. The new observation wells provided ground-water level, chloride concentration, hydraulic conductivity, and borehole geophysical data of the glacial aquifer. The glacial sediments range in thickness from less than 0.3 m to more than 76.2 m within the study area. A linear relation between Electromagnetic-induction (EM) conductivity log response and measured chloride concentration was determined. Using this relation, chloride concentration was estimated in parts of the glacial aquifer where sampling was not possible. EM logging is an effective tool to monitor changes in saltwater intrusion wedges.

**Keywords:** geophysics; groundwater; hydrogeology

---

## 1. Introduction

Manhattan Island is about 20.1 km long and 3.2 km wide and consists of unconsolidated deposits ranging from less than 0.3 m thick to more than 76 m thick overlying high-grade metamorphic bedrock in southern Manhattan (Figure 1). The lack of published water-table maps, unknown glacial aquifer characteristics, unknown extent of saltwater intrusion, and the need for a relation to estimate chloride concentration from EM conductivity log response prompted this research. The study area is south of Central Park (Figure 1). Manhattan is bounded on the west by the Hudson River, on the east by the East River, and on the south by New York Harbor (each of these tidal embayments contain saltwater). Industrial pumpage of the aquifer underlying southern Manhattan Island in New York City may have caused the chloride concentration to exceed the U.S. Environmental Protection Agency Secondary Maximum Contaminant Level (SMCL) of 250 mg/L in private-supply wells beginning in 1940–1950 and its effect was observed recently (2004–2006). Perlmutter and Arnow [1] indicate industrial pumpage was as high as 22.7 million liters per day (mld) in 1940–1950. Chloride concentrations in groundwater underlying southern Manhattan Island have remained unchanged, except in some areas where it has increased since the 1940s.

Electromagnetic-induction (EM) logging was completed on a set of observation wells along the southeastern coast of the study area to determine the thickness and extent of saltwater intrusion where screen samples were not possible. A set of polyvinyl-chloride (PVC) cased observation wells in nearby Long Island, New York was used to determine the relation between EM log response and sampled chloride concentration (Figure 2).

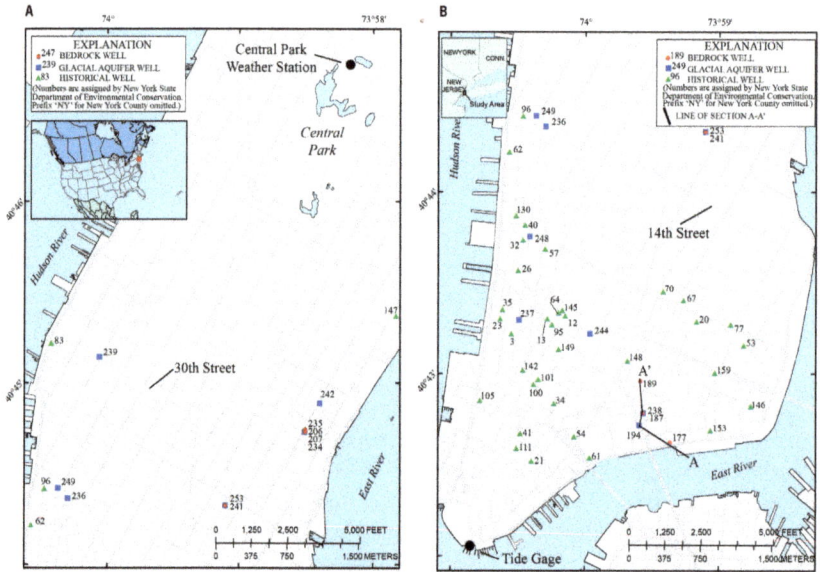

**Figure 1.** Locations of bedrock, glacial, and historical (glacial) wells within the (**A**) northern, and (**B**) southern parts of the study area in Manhattan, NY.

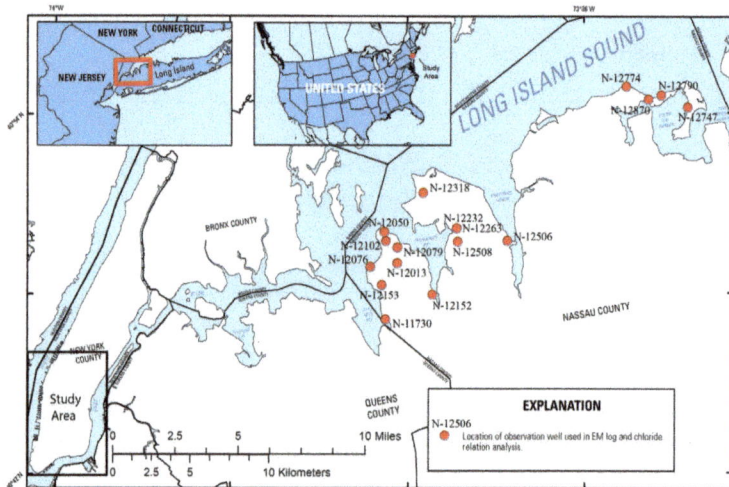

**Figure 2.** Locations of 16 observation wells with Electromagnetic induction (EM) conductivity logs and chloride concentrations used to calculate a relation between measured EM conductivity log response and chloride concentration in Nassau County, Long Island, NY. The southern Manhattan study area is shown.

## 2. Historical Background

In 1865, a topographic and hydrographic map of Manhattan Island was completed with the modern street grid superimposed [2]. The Viele [2] map shows flowing streams and wetland areas within the study area indicating a shallow depth to water in the unconsolidated aquifer system. Since that time, intensive urbanization has filled in and covered over most of the recharge areas and surface water expressions, severely reducing recharge to this groundwater flow system in Manhattan Island. Analysis using a Geographic Information System (GIS) indicates more than 90% impervious surfaces within the study area. Examination of extensive wetland areas and numerous flowing streams in the Viele [2] map suggests a much higher water table must have existed in 1865 before the installation of impervious surfaces in the early part of the 20th century. A geologic map with cross sections based upon test borings for several subway tunnels describes the unconsolidated sediments underlying Manhattan as consisting of sand and gravel deposits [3]. In 1905, a compilation of drill core data from various projects in Manhattan Island were used to map the elevation of the bedrock and describe the overlying unconsolidated sediments along several sections [4]. Hobbs describes the sediments in southern Manhattan as consisting of sand, fine sand, and gravels with some of the thickest deposits on Manhattan Island. Perlmutter and Arnow [1] describe the unconsolidated sediments in Manhattan Island as consisting of till and ground moraine in the northern part of the study area and stratified drift in the southern part of the study area. These deposits consist mainly of sand and gravel with localized deposits of silt and clay. Perlmutter and Arnow [1] also list the chloride concentrations of industrial wells screened in the glacial aquifer within the study area. Public-supply of drinking water is supplied via pressurized water tunnels beginning around 1910. The only published records of industrial pumping of the glacial aquifer are not very precise and appear to begin around the beginning of the 20th century and end sometime after 1953 [1]. It appears that within thirty years saltwater intrusion had occurred [1].

Recent drilling (2004–2006) indicates the saturated part of the glacial aquifer in southern Manhattan only extends as far north as about 30th Street, with small localized exceptions in stream channels (small buried valleys) of Pleistocene age to the north (Figure 1). A large deeply eroded channel in southern Manhattan Island dominates the study area and is infilled with saturated sediment that can produce significant quantities of groundwater [1]. Baskerville [5] included these Pleistocene stream channels in his engineering maps of the study area.

## 3. Methods

Data collected during this study included ground-water levels, chloride concentrations, and EM logs which are available at https://waterdata.usgs.gov/ny/nwis/gw, https://waterdata.usgs.gov/ny/nwis/qw, and https://webapps.usgs.gov/GeoLogLocator/, respectively. In addition, this study utilizes a nearby network of sixteen PVC cased wells on Long Island, New York to collect groundwater samples for chloride concentration measurement from well screens and core samples, and electromagnetic induction log responses from those zones to produce an equation that describes the relation between EM conductivity log response and chloride concentration (Figure 2). This equation was used to estimate chloride concentrations in several PVC cased wells within the study area that did not have screens and groundwater samples for chloride concentration measurement. The aquifers underlying Manhattan Island and nearby Long Island are similar (coastal plain aquifers) in that they are quartz-rich and have very low percentages of fine materials (silts and clays) [6].

### 3.1. Manhattan Island Well Network

Fourteen wells were used for the collection of ground-water levels, chloride-concentrations, aquifer-test data, and borehole geophysical logs (Figure 1, Table 1). Eleven of the fourteen wells, installed during 2004–2006 in southern Manhattan, were cased with PVC and screened in the glacial aquifer. Three of the eleven PVC wells were deep NX-sized (76 cm-diameter) wells (NY-177, NY-187, and NY-189) drilled by the mud-rotary method in the glacial aquifer and then drilled in bedrock by the diamond-core method to obtain continuous rock core samples (Figure 1, Table 1). These three wells had the unconsolidated sediment portion of the well cased off in PVC material with no screen. A fourth deep PVC cased and screened glacial well (NY-194) was also used (Figure 1). Geophysical log analyses were used on these wells to estimate chloride concentrations in the unconsolidated glacial aquifer. Additional lithologic and thickness data on the glacial aquifer were collected from drilling and gamma logs of 64 bedrock wells drilled throughout southern Manhattan [7].

**Table 1.** Well completion information and hydrogeologic data from observation wells referenced to NGVD 29.

| Hydrogeologic Unit | New York State ID Number | Land Surface Elevation (m) | Bottom Depth of Well (m) | Casing Length (m) | Water Level (m) above Mean Sea Level | Chloride in Glacial Aquifer (mg/L) | Hydraulic Conductivity (m/Day) | Transmissivity (m²/Day) |
|---|---|---|---|---|---|---|---|---|
| | NY 241 | 8 | 183 | 15 | 3.41 | NA | NA | NA |
| | NY 206 | 6 | 155 | 7 | NA | NA | NA | NA |
| | NY 207 | 6 | 183 | 8 | NA | NA | NA | NA |
| bedrock | NY 234 | 6 | 162 | 9 | NA | NA | NA | NA |
| | NY 189 | 13 | 183 | 71 | NA | NA | NA | NA |
| | NY 187 | 11 | 171 | 88 | NA | 11,800 ** | NA | NA |
| | NY 177 | 2 | 182 | 69 | NA | 7000 ** | NA | NA |
| | NY 239 | 6 | 8 | 6 | 1.16 | 641 | 6 | 48 |
| | NY 248 | 5 | 12 | 11 | −0.09 | 240 | 550 | 13,470 |
| | NY 237 | 5 | 12 | 11 | −0.04 | 1080 * | 12 | 297 |
| | NY 253 | 8 | 18 | 14 | 3.31 | 357 | NA | NA |
| | NY 235 | 6 | 8 | 6 | 0.95 | 1440 * | 0.05 | 0.10 |
| glacial aquifer | NY 242 | 11 | 20 | 18 | 4.47 | 405 | 0.04 | 0.20 |
| | NY 236 | 6 | 25 | 24 | 0.28 | 672 * | 39 | 660 |
| | NY 244 | 8 | 20 | 19 | 0.30 | 217 | 52 | 670 |
| | NY 194 | 9 | 65 | 62 | NA | 15,250 * | 0.9 | 53 |
| | NY 238 | 11 | 11 | 9 | 2.64 | 277 | 22 | 1660 |
| | NY 249 | 5 | 9 | 8 | 0.25 | 560 | 92 | 409 |

Notes: NA indicates the data are either not available or not applicable. * Indicates aquifer test grab samples analyzed with Cl probe. ** Indicates Cl value in the glacial aquifer estimated by EM log through PVC casing. Data available at http://waterdata.usgs.gov/nwis/si.

### 3.2. Groundwater Levels and Hydraulic Testing

Groundwater levels were measured during borehole-geophysical logging, water-level synoptics, and aquifer tests at the wells (Table 1). Two wells (NY-253 and NY-241) had a continuous digital recorder installed during the study to determine possible hydraulic interconnection with the underlying bedrock flow system at that location (Figure 1). Daily precipitation data were collected by the National Weather Service at the Central Park weather station (Figure 1) [8]. Tidal elevation data were collected by the National Oceanographic and Atmospheric Administration every six minutes at The Battery Park tidal station (Figure 1) [9]. A total of 10 wells had aquifer tests completed using a submersible pump that pumped the wells until the drawdowns stabilized at a constant rate for up to three hours. During the aquifer tests, drawdown was measured and the pumping rate was measured periodically using a calibrated bucket and stopwatch. The time-drawdown and pumping rate data were entered into a computer program for aquifer-test analysis [10].

The following analytical solutions were used for the analyses of the single-well aquifer test data for eight of the glacial aquifer wells: Neuman [11], Theis [12], and Moench [13]. The average hydraulic conductivity was estimated for each well using these methods (Table 1).

### 3.3. Chloride Concentrations

During the drilling of the sixteen Long Island observation wells, samples of groundwater were collected from core samples using the filter-press method using nitrogen gas to force pore-fluids from portions of core material that were uninvaded by drilling mud into sealed test tubes [14]. A calibrated chloride probe was inserted into the samples to obtain a chloride concentration in milligrams per liter. A Thermo-Scientific chloride ion probe calibrated to laboratory standard solutions with an accuracy of 10% was used to measure the concentration of chloride in the ground-water samples at room temperature [15]. Eleven of the Manhattan observation wells had been installed prior to this study and only pumping of groundwater through their screen zones were possible to obtain representative samples for chloride concentration measurement of the glacial aquifer. Pumping usually lasted from two to three hours after hydraulic equilibrium and field water quality parameter data indicated equilibrium was achieved, a representative sample was obtained according to protocols outlined by Koterba and others [16] and analyzed for major inorganic analysis including chloride concentrations at the USGS water quality laboratory [17] (Table 1).

*3.4. Borehole-Geophysical Logging*

Borehole-geophysical logs used for this study included natural gamma and focused electromagnetic induction (EM conductivity). The logging methods have been described by Archie, 1942, Keys and MacCary, 1971; Serra, 1984; Keys, 1990; McNeill and others, 1996; and Williams and Lane, 1998 [18–23]. Gamma logs were used for lithologic and stratigraphic correlation. Gamma log response is generally low in the quartz-rich sand aquifers of Pleistocene and Cretaceous age found within New York City and Long Island coastal plain deposits. The exception was the Raritan clay, a Cretaceous aged clay unit underlying nearby Long Island that exhibits significant gamma responses [6,24]. The unconsolidated sediments underlying Manhattan Island are of Pleistocene age based upon core analysis and previous work [1]. EM conductivity logs provided an electrical conductivity profile of the formation, from which ground-water conductivity and chloride concentrations can be inferred [25,26].

Electromagnetic induction (EM) conductivity logs were collected using a Geonics model EM-39 tool that employs coaxial coil geometry with an intercoil spacing of 50 cm to allow substantial radius of measurement into the formation with excellent vertical resolution [27,28]. Measurement is unaffected by conductive borehole fluid or the presence of plastic casing. The combination of a large conductivity range, high sensitivity and very low noise and drift, allows accurate measurement of subsurface conditions [27].

In the NYC and Long Island area the EM log responses for brackish to saltwater saturated materials are tens to hundreds of times higher than lithologic changes in the regional sediments [29]. EM logs were used to delineate saltwater intrusion in other earlier studies in Florida, and California [30,31]. In the California study, a relation between bulk EM resistivity and pore-fluid conductance was determined for predictive studies [32]. Within the New York City and Long Island regional area, the aquifers and groundwater are highly resistive, therefore EM conductivity log response is very sensitive to slight increases in groundwater conductivity due to increased dissolved solids [29]. A nearby network of sixteen PVC cased wells in the coastal plain sediments on Long Island, NY similar to those encountered on Manhattan Island were used to collect calibrated EM conductivity logs of sections where groundwater samples for chloride concentration measurement were obtained (Table S1 in Supplementary Materials). Gamma log and EM conductivity log responses are typically low in the region's aquifers due to the high quartz content (Figure 3). Well N-12506, one of the sixteen Long Island observation wells used in this study to produce the EM log response to chloride concentration relation, was used to show how gamma and EM conductivity logs relate to changes in chloride concentration and geology (Figure 3). The chloride concentrations and EM log response from sixteen Long Island observation wells were used to determine the relation between chloride concentration in milligrams per liter (mg/L) and measured EM log response in millisiemens per meter (mS/m) (Figure 4). The main purpose of using the relation between EM conductivity log response and chloride concentration is to allow the conversion of EM log response in wells where chloride sampling is not possible to estimate the chloride concentrations in the aquifer. A similar technique was used in California in PVC cased wells where groundwater samples for chloride concentration measurement were available with EM conductivity logs, and in Florida to map the saltwater interface using surface and borehole EM methods to produce logarithmic equations relating EM response to chloride concentration [25,33].

A linear relation was observed in the data from Long Island wells between EM conductivity log response and chloride concentration collected from screen zones in the well and pore fluid samples obtained from cores during drilling (filter-press) (Figure 4). A least-squares regression was developed to relate changes in EM conductivity to changes in chloride concentration in groundwater from 16 wells using 43 samples (Table S1):

$$Cl = 25.26 \, (EM) + 10.1 \qquad (1)$$

where Cl is the chloride concentration (milligrams per liter) in groundwater from screen zones of wells and filter press samples from cores, and EM is the peak electromagnetic conductivity (millisiemens per meter) from the EM conductivity geophysical log over the screen zone length (6 m) and core sample length.

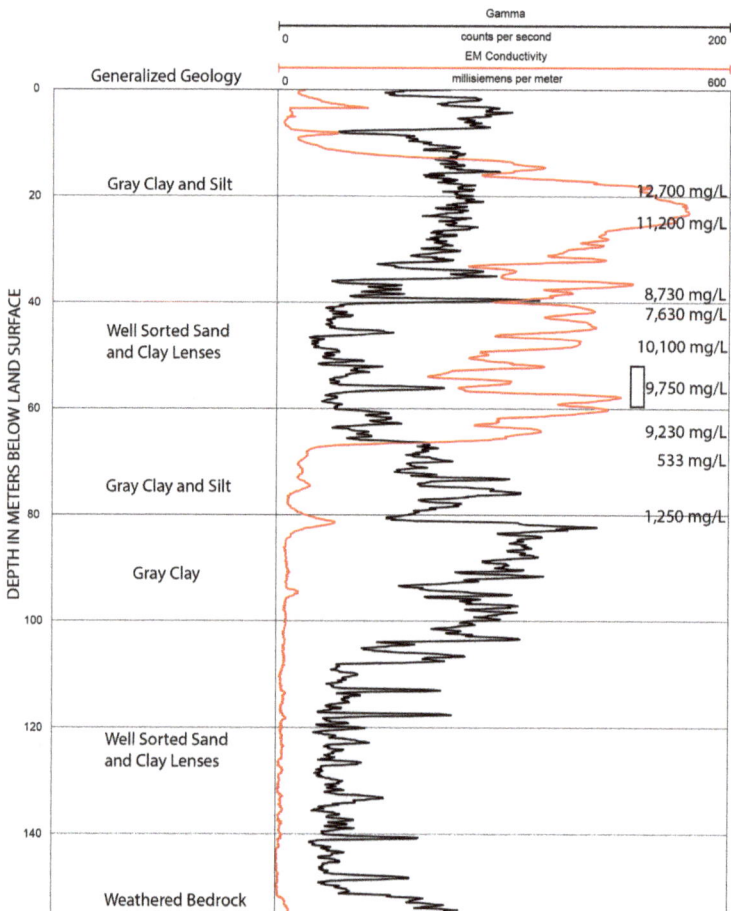

**Figure 3.** Generalized geology, natural gamma log, EM conductivity log, and chloride concentrations (milligrams per liter) in observation well N-12506 Long Island, NY (location shown in Figure 2).

The relation between these two variables is significant with an $R^2$ of 0.96 (Figure 4). A *p*-value of $<2.2 \times 10^{-16}$ indicates significant differences between the two datasets and a *t*-test value of

33.3 indicates a strong relation [34]. The reason for the strong correlation was due to the consistent EM conductivity log field calibration, and the region's aquifers that typically consist of high quartz content and low silt and clay concentrations providing minor EM conductivity response from the geology compared to the primary EM conductivity response that was very sensitive to minor increases in chloride concentrations of the groundwater.

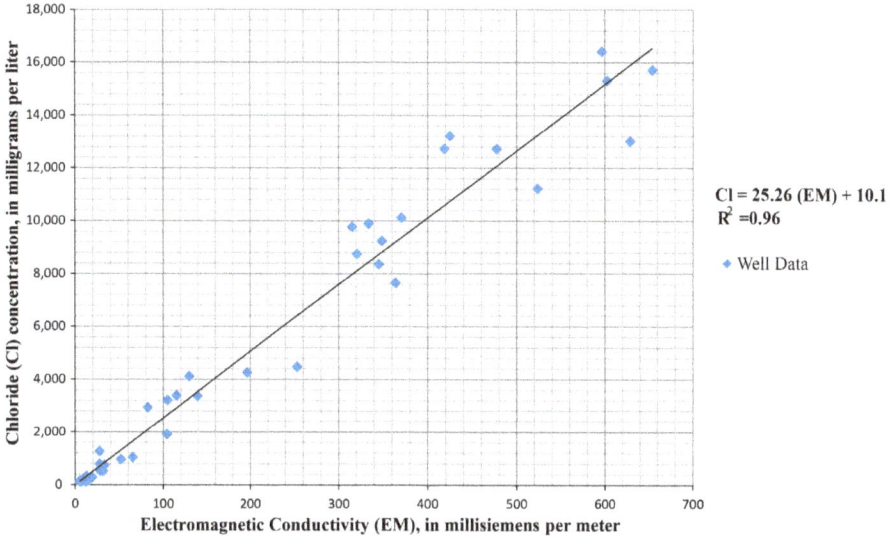

**Figure 4.** Chloride concentration (milligrams per liter) in ground water as a function of electromagnetic induction (EM) conductivity log response (millisiemens per meter) in 16 wells shown in Figure 2 in Nassau County, Long Island, NY.

## 4. Hydrogeology of the Glacial Aquifer in Manhattan Island

Analyses of drilling logs, ground-water levels, aquifer-test data, chloride concentrations, and borehole-geophysical logs provide new insight on the hydrogeologic properties of the glacial aquifer in southern Manhattan Island. The term glacial aquifer is introduced in this paper to represent a sequence of unconsolidated Pleistocene-age glacial sediments that overlie the bedrock in southern Manhattan. The sediments consist of gravel, sand, silt, and clay underlain by crystalline bedrock of Proterozoic and Paleozoic age [1,3,4,35–37].

### 4.1. Extent and Thickness of the Glacial Aquifer

The southern Manhattan study area is underlain by unconsolidated glacial drift and till deposits of Pleistocene age throughout most of the study area, except in a few areas in the northern part (near Central Park) where bedrock outcrops are present (Figure 1).

The topography of the northern part of the study area is dominated by the bedrock outcrops and a veneer of soils above them. Drilling data indicate the sediment thins to less than 6 m in a large part of the northern study area north of 30th Street (Figure 1). The sediment thickness increases to 15.2 m or more in very narrow buried valleys in the northern part of the study area (north of 30th Street). In contrast, an extensive buried valley containing over 76 m of sediment was delineated in most of the southernmost part of Manhattan [1,3,4,35,36]. Split-spoon core analysis of a monitoring well (NY-194) near the East River within the southern buried valley indicates stratified-drift deposits consisting of sand and gravel (Figure 1). The gravels are composed of sandstone and diabase transported by glacial action from New Jersey (Figure 1).

## 4.2. Water-Table Map of the Glacial Aquifer

This paper presents the first comprehensive water-table map constructed for the unconsolidated glacial aquifer in southern Manhattan (Figure 5). The presence of a clay unit was indicated in the southwestern-most part of Manhattan and may partially confine groundwater locally at depth. Water-level elevation (hydraulic head) ranged from 4.5 m above National Geodetic Vertical Datum of 1929 (NGVD 29) at NY-242 to −0.09 m (below NGVD 29) at NY-248 (Figure 5, Table 1). The highest water levels occur in the central part of the study area, which appears to be the recharge zone for the glacial aquifer (Figure 5). Groundwater flows from the central part of the study area toward the southern, western, and eastern coastal discharge zones. The ground-water divide in the glacial aquifer appears to follow the central topographic ridge of the island, but is slightly shifted towards the east probably due to the effect of dewatering operations in subway and rail stations located in the southwestern part of the study area (Figure 5). A cone of depression along the southwest part of the study area was indicated in the water-table map and includes wells NY-248 and NY-237 due to the dewatering of the glacial aquifer in the area for subway stations (Figures 1 and 5). The bedrock potentiometric-surface below this area is also depressed suggesting a possible hydraulic interconnection between the glacial aquifer and the bedrock in this area. Leakage from water mains and sewers probably contributes some recharge to the glacial aquifer which is mostly covered by impervious surfaces except in small neighborhood parks.

Water levels digitally recorded in two wells at E22nd Street in the central part of the study area (one screened at 14 m in the glacial aquifer (NY-253), and the other open in fractured-bedrock from 15.2 m to 183 m (NY-241)) indicate similar aquifer responses in both ground-water flow systems to precipitation events (Figures 1, 5 and 6). Both the glacial aquifer and bedrock water levels indicate this area is a recharge area. The glacial aquifer well (NY-253) shows a slight delay in recharge from precipitation as compared to the bedrock well (NY-241) (Figures 1 and 6). The bedrock aquifer well had a slightly higher water elevation (hydraulic head) (0.1 m) than the glacial aquifer well (Figure 6). The slightly higher water level elevation in the bedrock, the faster recharge response to precipitation compared the glacial aquifer at this location, and the large amount of impervious surfaces within the study area suggests the bedrock aquifer may be a source of recharge to the glacial aquifer in this part of Manhattan Island (Figures 1, 5 and 6). The glacial aquifer well (NY-253) does not show clear indications of a tidal influence while the bedrock well (NY-241) does show a small delayed influence (Figures 1 and 6).

Pumping of two bedrock wells, NY-207 and NY-234, along the eastern side of the study area during aquifer testing had no effect on water levels in an adjacent glacial aquifer well NY-235 (Figure 1) [7]. The glacial aquifer appears to be not hydraulically connected to the transmissive bedrock fractures in this area.

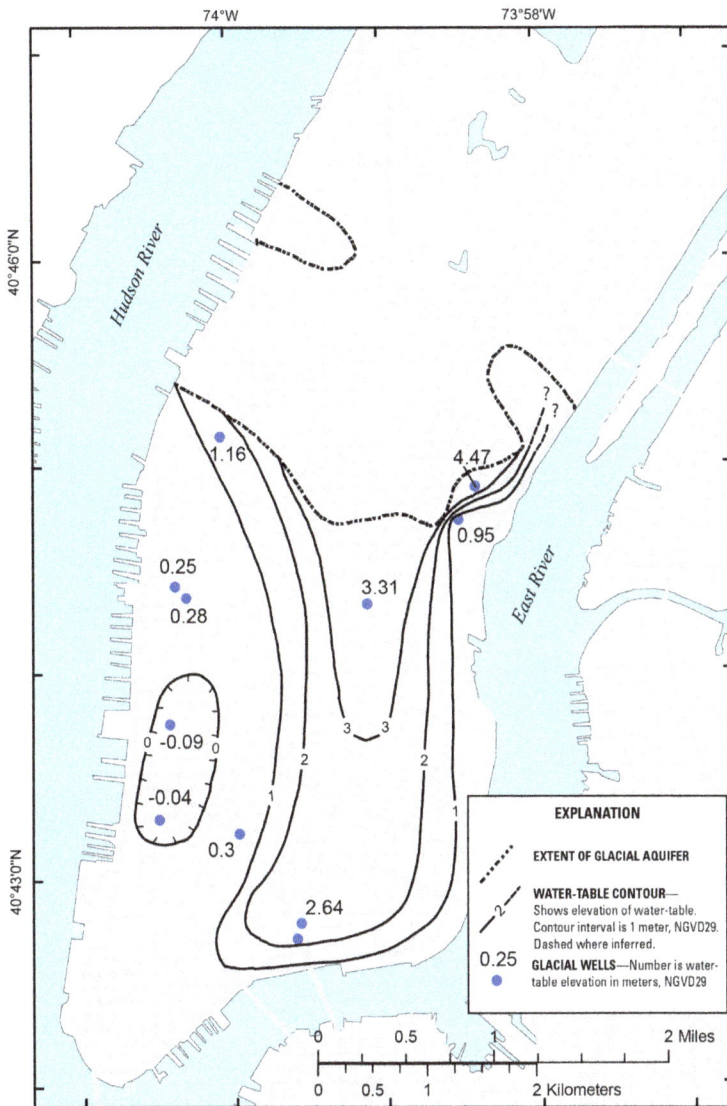

**Figure 5.** Composite water-table elevation map in the glacial aquifer in Manhattan Island, NY (2004–2006).

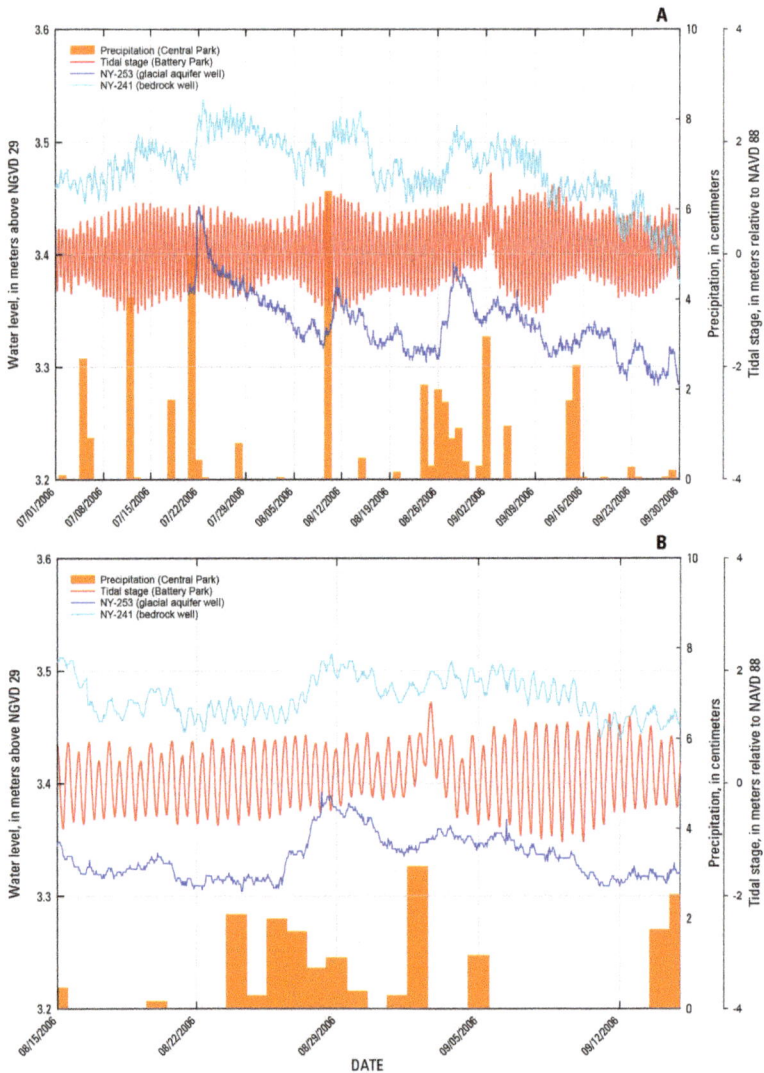

**Figure 6.** Precipitation, tidal data, and water levels in the fractured bedrock and glacial aquifer on the south part of Manhattan, NY for the periods of: (**A**) 1 July–30 September 2006; and (**B**) 15 August–15 September 2006 (locations of wells, precipitation and tidal stations shown in Figure 1).

*4.3. Hydraulic Properties of the Glacial Aquifer*

Aquifer tests of the 10 glacial aquifer wells indicate a wide range of aquifer hydraulic conductivity. The estimated hydraulic conductivity of the glacial aquifer ranges from 0.04 to 550 m per day at wells NY-242 and NY-248, respectively (Figure 1 and Table 1). The hydraulic conductivity data of the glacial aquifer in southern Manhattan indicate lower permeable sediments were present along the eastern extent of the aquifer, and the highest were measured in the western part of the study area. The highest values of hydraulic conductivity in the glacial aquifer were also present within the northwestern end of the deep southern buried valley.

The wide range in hydraulic conductivity may be due to the glacial origin of deposition with ground moraine and stratified drift.

## 5. Extent of Saltwater Intrusion of the Glacial Aquifer in Southern Manhattan Island

Industrial pumpage more than 70 years ago that may have caused significant saltwater intrusion into the glacial aquifer. On nearby Long Island, similar saltwater intrusion wedges were formed due to excessive pumpage of groundwater near coastal saltwater embayments [26,38]. Contouring of historical and current chloride concentration data indicates several wedges of saltwater intrusion in the study area. Borehole-geophysical logging using EM conductivity logs delineated the thickness and provided estimates of chloride concentrations in the most severely intruded saltwater wedge.

### 5.1. Chloride Concentrations in the Glacial Aquifer in 1940–1950

Perlmutter and Arnow [1] reported the location, depth, and chloride concentration of private wells on Manhattan Island, but did not contour the data. Examination of the historical data indicates 37 glacial aquifer wells used for private supply were sampled for chloride concentration from 1940 to 1950 within the study area (Figure 1). When these data are plotted, several wedges of saltwater intrusion are indicated in the first chloride concentration (isochlor map) of the glacial aquifer in southern Manhattan (Figure 7). Industrial pumpage in southern Manhattan during the 1940s–1950s was documented to be as high as 22.7 mld [1]. The limited recharge capability of this aquifer due to impervious surfaces and the extreme industrial pumpage created a severe stress on this system and is the probable mechanism for saltwater intrusion at that time.

**Figure 7.** Chloride concentrations in the glacial aquifer during 1940–1950, Manhattan Island, NY (saltwater wedges A, B and C shown) (chloride data [1]).

Three areas of historical saltwater intrusion are indicated and designated wedges A, B, and C (Figure 7). Historic saltwater wedge A in the southernmost part of the study area had a peak chloride concentration of 10,150 mg/L in well NY-61 and extended to well NY-101 (Figures 1 and 7). Saltwater wedge B is northwest of saltwater wedge A along the Hudson River coastline with a peak chloride concentration of 2800 mg/L at well NY-35. A third wedge of saltwater intrusion, wedge C, was indicated along the northwestern-most part of the glacial aquifer with a peak chloride concentration at well NY-83 of 2350 mg/L (Figures 1 and 7). Significant industrial pumpage from the glacial aquifer decreased after the 1950s [1].

*5.2. Present Chloride Concentrations in the Glacial Aquifer*

All of the wells sampled during 1940–1950 have either been destroyed or abandoned [1]. Samples of groundwater were collected from the 11 recently installed observation wells screened in the glacial aquifer during aquifer testing in 2004 and ground-water quality sampling in 2006. Chloride concentrations in the glacial aquifer range from 217 mg/L at well NY-244 to 15,250 mg/L at well NY-194 in 2004–2006 (Figures 1 and 8, Table 1). Most of the central part of the glacial aquifer appears to have chloride concentrations less than 100 mg/L. Four areas of saltwater intrusion are indicated and designated saltwater wedges A, B, C, and D (Figure 8).

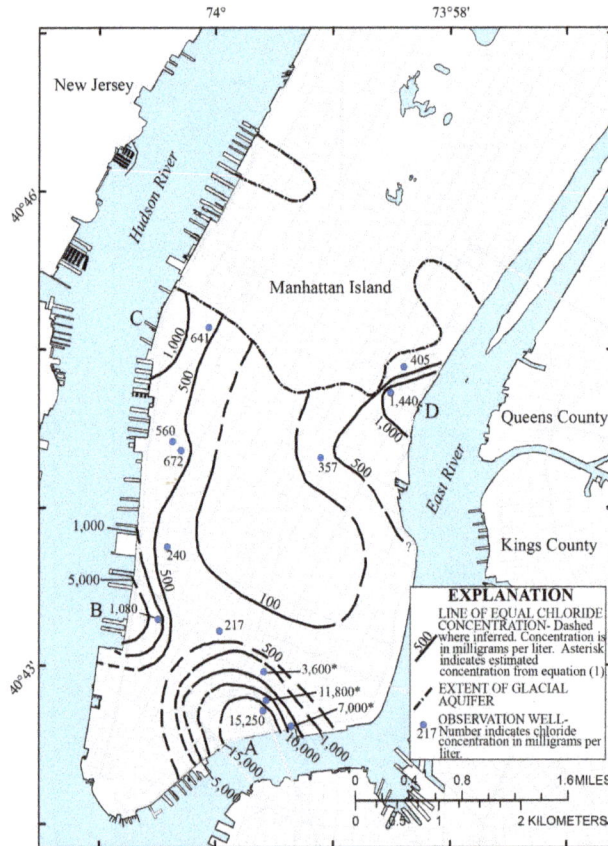

**Figure 8.** Chloride concentrations in the glacial aquifer during 2004–2006, Manhattan Island, NY (saltwater wedges A, B, C and D shown).

5.2.1. Saltwater Wedge A

EM logs of three bedrock wells (NY-177, NY-187, and NY-189) that had PVC casings in the unconsolidated overburden (glacial aquifer above bedrock) and one glacial aquifer well (NY-194) cased and screened in PVC facilitated the delineation of saltwater wedge A at the base of the glacial aquifer (Figures 1, 8 and 9). Similar methods were used to delineate the saltwater–freshwater interface on Long Island using EM conductivity logs [26,29]. EM conductivity logs of these wells indicate the highest conductivity was in wells NY-194 and NY-187 (Figures 1 and 9). Using Equation (1), the peak measured EM conductivity log responses were converted to estimated chloride concentrations in mg/L in the deeper parts of the glacial aquifer.

Well NY-194 had a sampled chloride concentration of 15,250 mg/L (Figures 1 and 9). The saltwater wedge is 25.6 m thick with a freshwater–saltwater interface at 27.7 m below land surface (−18.9 m elevation) at NY-194 (Figure 9). Further inland, at well NY-187, the wedge is 42 m thick, had a sharp (narrow) interface, and the peak EM log response was 470 mS/m, which correlates to an estimated chloride concentration of 11,800 mg/L using Equation (1) (Figures 1, 8 and 9). Adjacent to deep well NY-187 is a shallow well NY-238 with a depth of 11 m (Figures 1, 8 and 9). The water-table elevation at this well is 2.65 m above sea level which indicates an estimated depth of the saltwater interface of nearly 106 m below sea level under natural conditions using the Ghyben-Herzberg relation where the depth of the saltwater interface below sea level is about 40 times the freshwater elevation above sea level [39,40].

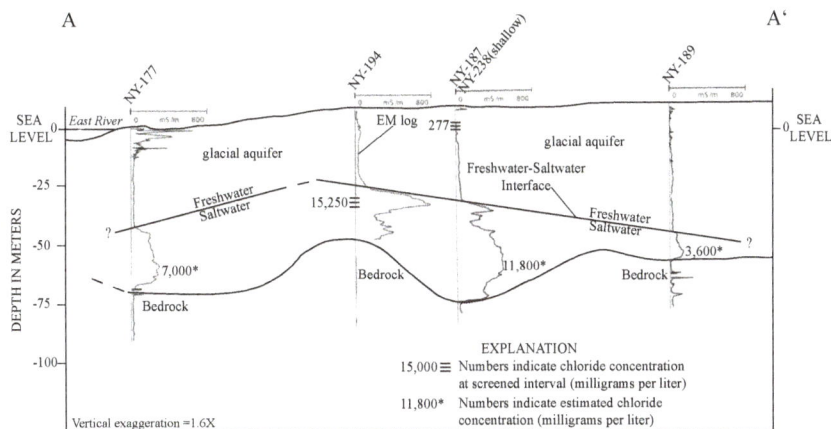

**Figure 9.** Hydrogeologic section showing the extent of saltwater wedge A, Manhattan Island, NY (location shown in Figures 1 and 8).

This supports the interpretation that the saltwater wedges in southern Manhattan Island were probably caused by industrial pumpage and are not natural. The sharp or narrow interfaces indicated by the EM logs are indicative of areas where active saltwater intrusion is due to pumpage, similar wedges are seen on nearby Long Island [26,29].

Well NY-177 is closest to the coastline; the saltwater wedge is 22.9 m thick, has a sharp (narrow) interface, and the peak EM response was 280 mS/m, which correlates to an estimated chloride concentration of 7000 mg/L (Figure 9). This indicates that this well is at the northeastern edge of saltwater wedge A (Figures 1, 8 and 9).

NY-189 was the farthest inland well where borehole EM logs were used (Figures 1 and 9). The saltwater wedge was 11.6 m thick with a sharp (narrow) interface and peak EM response was 143 mS/m, which correlated to an estimated peak chloride concentration of 3600 mg/L. This indicates that well NY-189 is at the toe of saltwater wedge A (Figures 1, 8 and 9). This wedge is a remnant of the historical saltwater wedge A indicated in 1940 to 1950 (Figures 7 and 8).

### 5.2.2. Saltwater Wedges B, C, and D Under Current Conditions

Saltwater wedge B is on the southwestern coastal part of the study area (Figure 8) and appears to be another remnant of a historical saltwater wedge documented in that area from 1940 to 1950 [4] (Figure 7). Well NY-237 is in this saltwater wedge and had a sampled chloride concentration of 1080 mg/L (Figures 1 and 8). Saltwater wedge C is a remnant of a historical saltwater wedge (Figures 1, 7 and 8). It is unclear if saltwater wedge C may be part of saltwater wedge B or a separate wedge unto itself. Another saltwater wedge D is indicated along the central-eastern coastline (Figure 8). Wells NY-235 and NY-242 had sampled chloride concentrations of 1440 and 405 mg/L, respectively. This wedge may have existed in 1940–1950 as indicated by well NY-147 with a chloride concentration of 223 mg/L (Figures 1, 7 and 8).

### 6. Discussion

The results of this study indicate a strong linear relation exists between bulk measured EM conductivity log response and chloride concentration in pore water of coastal plain aquifers. The use of this relation correlated very well with sampled results for saltwater that intruded coastal aquifers. The high correlation in the relation between chloride and EM log response suggests negligible influence of aquifer geology on the EM conductivity log response as opposed to elevated chloride concentrations from saltwater intrusion. Application of this relation was used to estimate chloride concentrations outside of PVC cased wells where no screen zones are present that would allow groundwater samples for chloride concentration measurement to be taken.

The existence of nearly identical saltwater intrusion wedges in southern Manhattan Island during 1940–1950 and 2004–2006 indicates little movement of the saltwater interface in over 70 years. EM log responses from several wells within one of the saltwater wedges indicate sharp interfaces. The limited recharge capability of the aquifer, due to impervious surfaces and the 22.7 mld of reported industrial pumpage early in the 20th Century was probably the cause for the saltwater intrusion into the glacial aquifer and the persistence of the historical saltwater intrusion wedges over time. A similar situation was observed in nearby Kings County on Long Island across the East River from the study area. EM logs collected in Kings County decades after the cessation of groundwater pumpage infer that limited recharge potential for the area due to urbanization and low water elevations allowed the persistence of historical saltwater intrusion wedges to be preserved [41].

### 7. Conclusions

The use of filter-press samples obtained from core samples and screen zone groundwater samples to measure chloride concentrations indicate good agreement between to these two methods within the margin of instrument drift. Using a nearby network of sixteen PVC cased wells, forty-three data points were used to generate a relation equation between EM log response and chloride concentrations. The equation has a strong correlation coefficient. Application of the relation equation will provide valuable estimation of the chloride concentration in this coastal plain aquifer where no screen zone exists and will help delineate zones of chloride concentrations that exceed the SMCL. In contrast to screen zone water samples that represent only a small portion of a borehole, EM logs can provide a complete cross-sectional measurement of the variation in saltwater intrusion with depth in a PVC cased well. The equation was used to estimate the chloride concentration within saltwater intrusion wedges in southern Manhattan Island.

Analysis of the geophysical logs, chloride concentration, and aquifer test data provided information on the glacial aquifer in southern Manhattan. The water-table map and water-level data for the glacial aquifer indicate a recharge zone in the central-northern part of the study area with discharge zones along the southern, western, and eastern coastal areas.

Historical chloride data from 1940 to 1950 in 37 glacial aquifer wells used for private supply were contoured. Three wedges of saltwater intrusion were indicated in the glacial aquifer at that time. The highest chloride concentrations were present in saltwater wedge A in the southernmost part of the study area.

During 2004–2006, chloride concentrations from 14 glacial aquifer wells indicated four saltwater wedges. Three of the saltwater wedges, designated A, B, and C, are probably remnants of the historical saltwater intrusion wedges delineated earlier. Current chloride concentrations in the glacial aquifer range from 217 mg/L in well NY-244 to 15,250 mg/L in well NY-194. EM logs of three bedrock wells (NY-177, NY-187, and NY-189) and one glacial aquifer well (NY-194) cased in PVC were used to delineate saltwater wedge A in the glacial aquifer.

EM logs provided a unique opportunity to delineate the thickness and concentration of saltwater wedge A. The EM logs indicate the freshwater–saltwater interface is still very sharp. Despite the cessation of industrial pumpage more than 70 years ago, the freshwater flow in the glacial aquifer has been unable to push the saltwater interface to any measurable degree back to its source. The limited recharge capability of the aquifer, due to impervious surfaces and the 22.7 mld of reported industrial pumpage early in the 20th Century was probably the cause for the saltwater intrusion and the persistence of the historical saltwater intrusion wedges over time. This suggests that the glacial aquifer has had only a limited recovery from the past industrial pumpage.

**Supplementary Materials:** The following are available online at www.mdpi.com/2073-4441/9/9/631/s1, Table S1: Peak EM log response and chloride concentration values for wells in Nassau County, NY.

**Acknowledgments:** The authors thank the personnel of the New York City Department of Environmental Protection and the New York City Department of Design and Construction for technical assistance, support, and access to their records and wells. The authors acknowledge the technical assistance of Anthony Chu and others of the USGS New York Water Science Center in Coram, New York. Any use of trade, firm, or product names is for descriptive purposes only and does not imply endorsement by the U.S. Government.

**Author Contributions:** Frederick Stumm conceived and designed the study; Frederick Stumm collected and analyzed field data; Michael D. Como analyzed aquifer test data; Frederick Stumm wrote the paper with the assistance of Michael D. Como.

**Conflicts of Interest:** The authors declare no conflict of interest.

## References

1. Perlmutter, N.M.; Arnow, T. *Ground Water in Bronx, New York, and Richmond Counties with Summary Data on Kings and Queens Counties, New York City, New York*; Bulletin GW-32; New York State Water Power and Control Commission: Albany, NY, USA, 1953; 86p.
2. Viele, E.L. *Sanitary and Topographical Map of the City and Island of New York*; NYC Council of Hygiene and Public Health, 1 Sheet, Scale 1:10,000; Ferd. Mayer & Co., Lithographers: New York, NY, USA, 1865.
3. Graether, L.F. *Geologic Map and Sections of Manhattan Island, New York*; 2 Sheets, Scale 1:20,000; Leonard F. Graether: New York, NY, USA, 1898.
4. Hobbs, W.H. *The Configuration of the Rock Floor of Greater New York*; U.S. Geological Survey Bulletin 270; US Government Printing Office: Washington, DC, USA, 1905; 96p.
5. Baskerville, C.A. *Bedrock and Engineering Geologic Maps of New York County and Parts of Kings and Queens Counties, New York*; U.S. Geological Survey Miscellaneous Investigations Series, Map I-2306, 2 Sheets, Scale 1:24,000; U.S. Geological Survey: Reston, VA, USA, 1994.
6. Suter, R.; de Laguna, W.; Perlmutter, N.M. *Mapping of Geologic Formations and Aquifers of Long Island, New York*; Groundwater Bulletin 18; New York State Water Power and Control Commission: Albany, NY, USA, 1949; 212p.
7. Stumm, F.; Chu, A.; Monti, J., Jr. *Delineation of Faults, Fractures, Foliation, and Ground-Water-Flow Zones in Fractured-Rock, on the Southern part of Manhattan, New York, through Use of Advanced Borehole Geophysical Techniques*; U.S. Geological Survey Open-File Report 04-1232; USGS New York Water Science Center: Reston, VA, USA, 2004; 212p.

8. National Oceanic and Atmospheric Administration. Monthly and Annual Precipitation at Central Park, Manhattan, New York. 2010. Available online: https://www.ncdc.noaa.gov/IPS/lcd/lcd.html?_page=1& state=NY&stationID=94728&_target2=Next+%3E (accessed on 23 August 2017).

9. National Oceanic and Atmospheric Administration. Tides and Water Levels at the Battery, Manhattan, New York. 2017. Available online: https://tidesandcurrents.noaa.gov/waterlevels.html?id=8518750 (accessed on 21 August 2017).

10. Duffield, G.M. *AQTESOLV for Windows Version 4.5 User's Guide*; HydroSOLVE, Inc.: Reston, VA, USA, 2007.

11. Neuman, S.P. Effect of partial penetration on flow in unconfined aquifers considering delayed gravity response. *Water Resour. Res.* **1974**, *10*, 303–312. [CrossRef]

12. Theis, C.V. The relation between the lowering of the piezometric surface and the rate and duration of discharge of a borehole using groundwater storage. *Eos Trans. Am. Geophys. Union* **1935**, *16*, 519–524. [CrossRef]

13. Moench, A.F. Flow to a borehole of finite diameter in a homogeneous, anisotropic water-table aquifer. *Water Resour. Res.* **1997**, *33*, 1397–1407. [CrossRef]

14. Lusczynski, N.J. *Filter-Press Method of Extracting Water Samples for Chloride Analysis*; Water-Supply Paper 1544-A; U.S. Geological Survey: Washington, DC, USA, 1961; 13p.

15. Thermo Fisher Scientific. *User Guide—Chloride Ion Selective Electrode*; Thermo FisherScientific Inc.: Waltham, MA, USA, 2008; 39p.

16. Koterba, M.T.; Wide, F.D.; Lapham, W.W. *Groundwater Data-Collection Protocols and Procedures for the National Water-Quality Assessment Program—Collection and Documentation of Water-Quality Samples and Related Data*; Open-File Report 95-399; U.S. Geological Survey: Reston, VA, USA, 1995; 113p.

17. Fishman, M.L. *Methods of Analysis by the U.S. Geological Survey National Water Quality Laboratory—Determination of Inorganic and Organic Constituents in Water and Fluvial Sediments*; Open File Report 93-125; U.S. Geological Survey: Reston, VA, USA, 1993; 217p.

18. Archie, G.E. The electrical resistivity log as an aid in determining some reservoir characteristics. *Trans. AIME* **1942**, *146*, 54–62. [CrossRef]

19. Keys, W.S.; MacCary, L.M. Application of borehole geophysics to water-resources investigations. In *U.S. Geological Survey Techniques of Water-Resources Investigations*; U.S. Geological Survey: Washington, DC, USA, 1971; 126p.

20. Serra, O. *Fundamentals of Well-Log Interpretation*; Elsevier: New York, NY, USA, 1984; 423p.

21. Keys, W.S. Borehole geophysics applied to water-resources investigations. In *U.S. Geological Survey Techniques of Water-Resources Investigations*; U.S. Geological Survey: Reston, VA, USA, 1990; 150p.

22. McNeill, J.D.; Hunter, J.A.; Bosner, M. Application of a borehole induction magnetic susceptibility logger to shallow lithological mapping. *J. Environ. Eng. Geophys.* **1996**, *1*, 77–90. [CrossRef]

23. Williams, J.H.; Lane, J.W. *Advances in Borehole Geophysics for Ground-Water Investigations*; Fact Sheet 002-98; U.S. Geological Survey: Reston, VA, USA, 1998; 4p.

24. Buxton, H.T.; Soren, J.; Posner, A.; Shernoff, P.K. *Reconnaissance of the Ground-Water Resources of Kings and Queens Counties, New York*; Open-File Report 81-1186; U.S. Geological Survey: Reston, VA, USA, 1981; 64p.

25. Metzger, L.F.; Izbicki, J.A. Electromagnetic-induction logging to monitor changing chloride concentrations. *Ground Water* **2013**, *51*, 108–121. [CrossRef] [PubMed]

26. Stumm, F. *Hydrogeology and Extent of Saltwater Intrusion of the Great Neck Peninsula, Great Neck, Long Island, New York*; U.S. Geological Survey Water-Resources Investigations Report 99-4280; U.S. Geological Survey: Reston, VA, USA, 2001; 41p.

27. Taylor, K.C.; Hess, J.W.; Mazzela, A. Field evaluation of a slim-hole borehole induction tool. *Ground Water Monit. Remediat.* **1989**, *9*, 100–104. [CrossRef]

28. McNeill, J.D. *Technical Note 20 Geonics EM39 Borehole Conductivity Meter Theory of Operation*; Geonics Limited: Mississauga, ON, Canada, 1986; 18p.

29. Stumm, F. Use of focused electromagnetic induction borehole geophysics to delineate the saltwater-freshwater interface in Great Neck, Long Island, New York. In *Symposium on the Application of Geophysics to Engineering and Environmental Problems*; Bell, R.S., Lepper, C.M., Eds.; Environmental and Engineering Geophysical Society: San Diego, CA, USA, 1993; Volume 2, pp. 513–525.

30. Paillet, F.; Hite, L.; Carlson, M. Integrating surface and borehole geophysics in ground water studies—An example using electromagnetic soundings in South Florida. *J. Environ. Eng. Geophys.* **1999**, *4*, 45–55. [CrossRef]

31. Hanson, R.T. *Geohydrologic Framework of Recharge and Seawater Intrusion in the Pajaro Valley, Santa Cruz and Monterey Counties, California*; U.S. Geological Survey Water-Resources Investigations Report 03-4096. Available online: http://pubs.usgs.gov/wri/wri034096 (accessed on 21 August 2017).

32. Land, M.; Reichard, E.G.; Crawford, S.M.; Everett, R.R.; Newhouse, M.W.; Williams, C.F. Ground-Water Quality of Coastal Aquifer Systems in the West Coast Basin, Los Angeles County, California, 1999–2002. U.S. Geological Survey Scientific Investigations Report 2004-5067. Available online: http://pubs.usgs.gov/sir/2004/5067 (accessed on 21 August 2017).

33. Fitterman, D.V.; Deszcz-Pan, M. Geophysical mapping of saltwater intrusion in Everglades National Park. In Proceedings of the 3rd International Symposium on Ecohydraulics, Salt Lake, UT, USA, 12–16 July 1999.

34. Zuwaylif, F.H. *General Applied Statistics*; Addison-Wesley: Boston, MA, USA, 1979; 434p.

35. Kemp, J.F. The geology of Manhattan Island [N.Y.]. *N. Y. Acad. Sci. Trans.* **1887**, *7*, 49–64.

36. Murphy, J.J.; Fluhr, T.W. *The Subsoil and Bedrock of the Borough of Manhattan as Related to Foundations*; Fourth Issue, Paper 212; Municipal Engineers Journal: New York City, NY, USA, 1944; pp. 119–157.

37. Merguerian, C.; Baskerville, C.A. The geology of Manhattan Island and the Bronx, New York City, New York. In *Northeastern Section of the Geological Society of America, Centennial Fieldguide*; Roy, D.C., Ed.; The Geological Society of America, Inc.: Boulder, CO, USA, 1987; Volume 5, 481p.

38. Lusczynski, N.J.; Swarzenski, W.V. *Salt-Water Encroachment in Southern Nassau and Southeastern Queens Counties, Long Island, New York*; Water-Supply Paper 1613-F; U.S. Geological Survey: Washington, DC, USA, 1966; 76p.

39. Ghyben, B.W. Nota in verband met de voorgenomen putboring nabij Amsterdam. *Tijdschr. Kon. Inst. Ing.* **1888**, *9*, 8–22.

40. Herzberg, A. *Die Wasserversorgung Einiger Nordseebäder*; J. Gasbeleucht. Wasserversorg: München, Germany, 1901; Volume 44, pp. 815–819, 842–844.

41. Chu, A.; Stumm, F. Delineation of the saltwater-freshwater interface at selected locations in Kings and Queens Counties, Long Island, New York, through use of borehole geophysical techniques. In Proceedings of the 2nd Geology of Long Island and Metropolitan New York Conference, Stony Brook, NY, USA, 15 April 1995.

*water*

MDPI

*Article*

# The Interrelations between a Multi-Layered Coastal Aquifer, a Surface Reservoir (Fish Ponds), and the Sea

Adi Tal [1,2,*], Yishai Weinstein [2], Stuart Wollman [3], Mark Goldman [4] and Yoseph Yechieli [3,5]

[1]    Hydrological Service, Jerusalem 91360, Israel
[2]    Department of Geography and Environment, Bar Ilan University, Ramat Gan 5200100, Israel;
       Yishai.Weinstein@biu.ac.il
[3]    Geological Survey of Israel, Jerusalem 95501, Israel; swollman@gsi.gov.il (S.W.); yechieli@gsi.gov.il (Y.Y.)
[4]    The Geophysical Institute of Israel, Lod 7019802, Israel; mgol1302@gmail.com
[5]    Department of Environmental Hydrology & Microbiology (EHM), Ben-Gurion University,
       Sede Boqer 8499000, Israel
*    Correspondence: adit20@water.gov.il

Received: 1 September 2018; Accepted: 3 October 2018; Published: 11 October 2018

**Abstract:** This research examines the interrelations in a complex hydrogeological system, consisting of a multi-layered coastal aquifer, the sea, and a surface reservoir (fish ponds) and the importance of the specific connection between the aquifer and the sea. The paper combines offshore geophysical surveys (CHIRP) and on land TDEM (Time Domain Electro Magnetic), together with hydrological measurements and numerical simulation. The Quaternary aquifer at the southern Carmel plain is sub-divided into three units, a sandy phreatic unit, and two calcareous sandstone ('Kurkar') confined units. The salinity in the different units is affected by their connection with the sea. We show that differences in the seaward extent of its clayey roof, as illustrated in the CHIRP survey, result in a varying extent of seawater intrusion due to pumping from the confined units. FEFLOW simulations indicate that the FSI (Fresh Saline water Interface) reached the coastline just a few years after pumping has begun, where the roof terminates ~100 m from shore, while no seawater intrusion occurred in an area where the roof is continuous farther offshore. This was found to be consistent with borehole observations and TDEM data from our study sites. The water level in the coastal aquifer was generally stable with surprisingly no indication for significant seawater intrusion although the aquifer is extensively pumped very close to shore. This is explained by contribution from the underlying Late Cretaceous aquifer, which increased with the pumping rate, as is also indicated by the numerical simulations.

**Keywords:** seawater intrusion; multi-layered coastal aquifer; offshore geophysics; numerical model; tidal signal; sea–aquifer relations; fish ponds

## 1. Introduction

The extensive rural and urban development in coastal areas strongly impacts the coastal environment. This partly occurs via subterranean interaction between land and the sea, which involves a bi-directional flow, including seawater intrusion [1,2] and Submarine Groundwater Discharge (SGD, [3–5]). The latter may cause deterioration in water quality of the shallow sea, leading to algae blooms [6–9]. In some coastal areas, there are surface reservoirs (e.g., fish ponds) above the aquifer which can make the hydrogeological conditions more complex. Worldwide aquaculture has been annually increasing by 8.7% over the past 40 years, the fastest-growing food-producing sector [10]. One of the key environmental concerns about aquaculture is water degradation, due to the discharge of effluents with high levels of nutrients and suspended solids into adjacent waters, causing eutrophication, oxygen depletion and siltation [11]. In the study area, it was shown that the phreatic unit beneath the fishponds area has indeed been deteriorating. It is rich in phosphate, ammonium, Total Organic Carbon (TOC) and manganese, and characterized by strong reducing conditions [12].

Seawater intrusion (SI) is also a global concern, exacerbated by increasing demands for freshwater in coastal zones and predisposed to the influences of rising sea levels and changing climates. Open questions still exist, in particular about the transient SI processes and timeframes, and the characterization and prediction of freshwater–saltwater interfaces over regional scales and in highly heterogeneous and dynamic settings [13].

Numerical solutions have become standard tools for analyzing aquatic systems, which involve a bi-directional flow [14]. Two different modeling approaches have been used in the literature to represent the freshwater–saltwater relationship, one of which is sharp interface approximation [15–17], while the other includes a transition zone between freshwater and seawater [18–21]. Numerical solution with the FEFLOW cod is a common tool to study the salt water interface movement [22]. Yet, extensive measurement campaigns are necessary to accurately delineate interfaces and their displacement in response to real-world coastal aquifer stresses, encompassing a range of geological and hydrological settings [13]. Analytical studies of tide-induced groundwater fluctuations in confined coastal aquifers that extend under the sea have been studied in References [23–25] and others. Other studies [24,26] suggested a general analytical solution that showed that the tidal fluctuation in an inland borehole decreases with the confined layer roof length under the sea until a certain distance offshore, wherefrom it remains constant due to the loading effect.

The direct discharge of groundwater to the sea was quantitatively studied by several authors [27–29]. A common way of assessing the discharge to the sea is the use of numerical simulations and analytical solutions of the water flow regime [30,31]. This is aided by geophysical [32] and the use of chemical and isotopic tracers, including radon and radium isotopes [5,6,33]. Seepage meters, which are deployed on the seabed, provide direct local measurements of fluxes [34].

The objective of this study was to investigate the effect of the marine and terrestrial boundary conditions on the hydraulic connections between the sea and a multi-layered coastal aquifer and its resultant seawater intrusion, using numerical simulations and field observations.

## 2. Hydrogeological Background of the Study Area

The study area extends from Mt. Carmel to the sea at the southern Carmel coastal plain, Israel (Figure 1). It is covered by Quaternary Nilotic sands with a total thickness of about 40 m, which are underlain by Late Cretaceous carbonate basement [35]. Located on its eastern part are the Taninim springs at +3 m asl, which are the northern discharge of the Late Cretaceous Yarkon–Taninim carbonate Aquifer (Figure 1). These springs are the principal source of water for fishponds in this area. The spring water is brackish (1500–2000 mg $L^{-1}$ Cl), which is due to the mixing of fresh groundwater with saline water that exists in the western part of the Late Cretaceous aquifer [36]. The Late Cretaceous carbonates change from limestone and dolomite facies in the Mt. Carmel area on its eastern side to a chalky facies on its western (seawards) side [35,37], which prevents the direct flow of groundwater from the Late Cretaceous aquifer to the sea and forces the water to discharge in the Taninim springs through Quaternary sediments (Figure 3). Some of this water flows westward through the Pleistocene sediments and plays an important role in the water balance of the coastal Quaternary Aquifer [12,38–40].

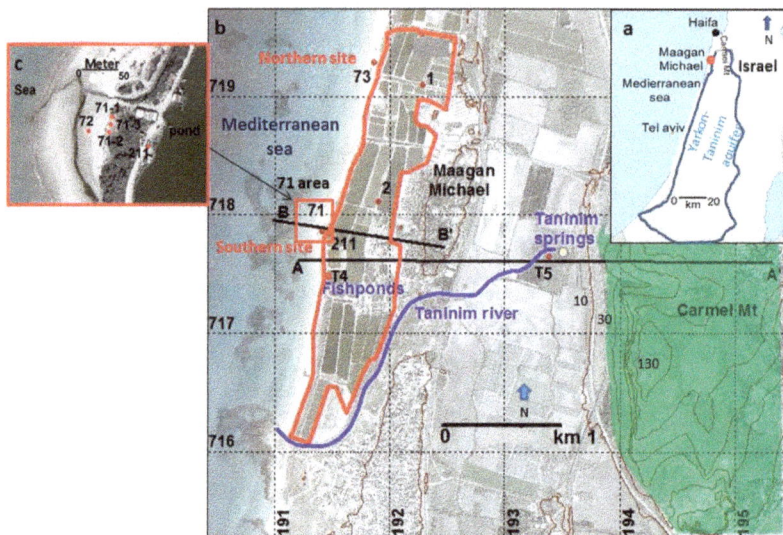

**Figure 1.** (**a**) Location map of the research area. (**b**) Borehole locations are shown as red points. Ponds' area is circled by a Red line. The area where the Cretaceous aquifer is exposed is colored green, and the Quaternary coastal aquifer is grey. (**c**) The left inset is an aerial photo of wells 71 area. The hydrogeological sections presented in Figures 3 and 2a were taken along the lines A–A′ and B–B′.

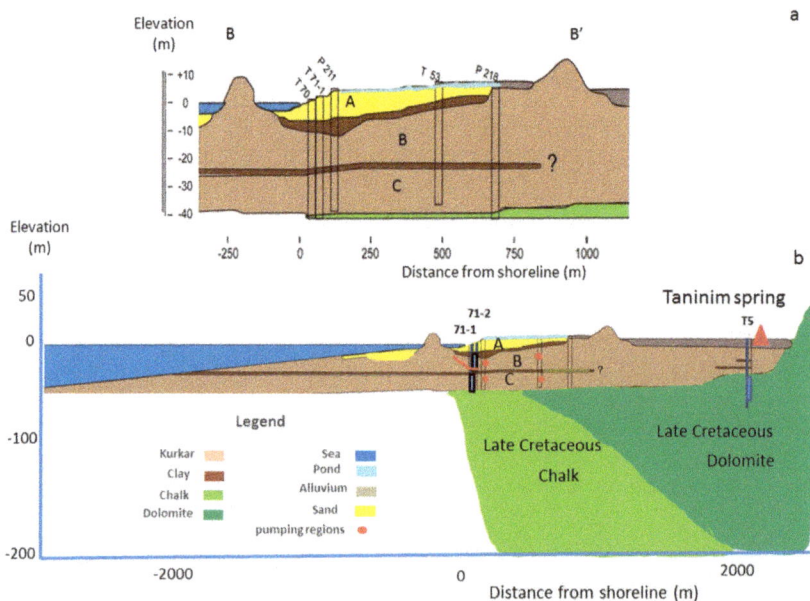

**Figure 2.** (**a**) Detailed hydrogeological section of the coastal aquifer in the study area (B–B′ in Figure 1). The coastal aquifer is subdivided into three units which are separated by clay layers: upper phreatic sandy unit (A) and two confined calcareous sand (Kurkar) units (B and C). (**b**) A hydrogeological cross section in the southern part of the studied area. The eastern boundary of the model is not shown (4 km from the shoreline).

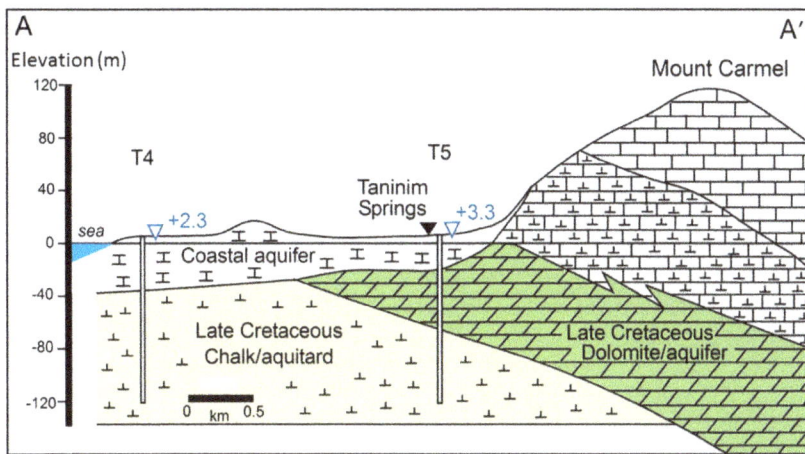

**Figure 3.** Hydrogeological section in the research area (A–A′ in Figure 1), modified after [35]. The coastal aquifer overlays Late Cretaceous limestone and dolomite aquifer in the east and chalk aquitard in the west. The dolomite aquifer (green) is the main source to the Taninim springs. The upper units in the Mount Carmel area, consisting of limestone and chalky limestone, are also of Late Cretaceous age. The blue number indicates the pre-utilization heads.

The thickness of the coastal Quaternary aquifer in the study area is 40 m (Figure 2a). It includes three sub-aquifers (Figure 2a). The shallow sub-aquifer is a phreatic Holocene sandy unit (unit A) with a thickness of several meters, which is separated from the underlying Pleistocene calcareous sandstone (locally called Kurkar) aquifer by a clay unit. The Pleistocene Kurkar is subdivided by another clay layer at an approximate depth of 20 m into two confined units, B and C, which in turn are underlain by the Late Cretaceous chalks. The coastal area is characterized by two N–S oriented Kurkar ridges on its western side, with the western one often partly or completely submerged [41,42]. In the study area, fish ponds occupy the area between the eastern ridge and the sea (Figures 1 and 2a), with an annual infiltration to the aquifer of about $9 \times 10^6$ m$^3$year$^{-1}$ [40].

The confined units B and C, which contain brackish water (1500–2000 mg L$^{-1}$ Cl), have been extensively exploited since 2005 as raw water for a desalination plant and for the fish ponds. The pumping in this area currently reaches ~$20 \times 10^6$ m$^3$year$^{-1}$. Although exploitation is very high, groundwater level in these units is quite stable, and seawater intrusion is very limited [43]. Geochemical mass balance indicates that more than 85% of the water in units B and C derives from the underlying Late Cretaceous aquifer [12], while the ponds contribution is no more than $3 \times 10^6$ m$^3$year$^{-1}$ [12]. This suggests that most of the ponds' losses flow through unit A to the sea.

## 3. Field Methods

The relations between the aquifer and the sea were studied in two coastal sites, 1.5 km apart (Figure 1). Groundwater level was mainly studied in boreholes drilled at 45–70 m from the sea to a depth of 10–40 m (wells 71-1, 71-2, 71-3, 72, 73, Figure 1), each opened to a different unit, A, B or C. We also used data from a few observation boreholes located farther inland (1500–600 m from the sea).

TDEM (Time Domain Electro Magnetic) measurements were conducted to locate the fresh–saline water interface. Surveys in coastal aquifers [44–47] show that this method is feasible for detecting the interface due to the very high resistivity contrast between the sea water saturated portion of the aquifer (1–2 ohm-m) and the fresh water part, which is characterized by values more than 10 ohm-m [45]. The survey was conducted using the Geonics EM-67 TDEM system (GEONICS Ltd., Mississauga, ON, Canada). The TDEM method utilizes a step-wise current flowing in a squared transmitter loop that produces a transient secondary electromagnetic field in the subsurface. Depending on the required exploration depth, the loop size varies between tens by tens meters to hundreds by hundreds meters. In the described survey, the loop size was 25 × 25 m. The electromagnetic field induced into the ground produces a secondary response, which is picked up on the surface by a receiver coil. The data were analyzed using two different inversion methods, smooth inversion, and sharp boundary inversion. The combination of solutions usually improves both the accuracy and reliability of decryption. The TDEM survey was conducted along two E–W traverses, one in the southern site and the second in the northern site (Figure 1). In each profile, three loop measurements were carried out between the shore and approximately 80 m inland. The TDEM survey and the interpretation of the data were conducted by the Geophysical Institute of Israel.

CHIRP sub bottom profiler was used for mapping of the shallow sub-seafloor sediments. The seismic survey was conducted with the "Adva" boat of the IOLR (Israeli Oceanography and Limnology Research), which also carried a Trimble AgGPS 132–receiver, an Odom Echotrack MKIII depth meter, and Oceanographer navigator. The depth meter was calibrated by Bar check and sound wave velocity meter in water, SV-2000 Smart sensor, which belongs to AML company (Hong Kong, China). The velocity of sound waves in the submarine sand and clay was taken as 1700 m/s, while for the sea water, it is 1530 m/s [48]. The CHIRP survey and the interpretation were conducted by Arik Golan from IOLR.

Electric conductivity (EC) profiles were conducted in the monitoring wells 71-1, 71-2, and 71-3 (locations in Figure 1) using a Robertson Geologging (RG) profiler.

Water level was measured by Schlumberger divers with 5 min frequency in wells 1, 2, 71-1-3, and 73. Sea Level was measured at Hadera MedGloss station, 2.5 km from the shore by IOLR (Israeli Oceanography and Limnology Research).

Slug tests were conducted for the determination of the hydrological conductivity in the aquifer. In each well, we conducted five tests, including three nitrogen injections, slug in and slug out. The slug tests were conducted in the monitoring wells 1, 2, 71-1, 71-2, 72, and 71-3. Results are presented in the Supplementary Materials, while the interpretation follows the procedure given in [49].

Point dilution test with Uranine solution was implemented in well 71-3, which penetrates the shallow phreatic aquifer. The tracer, with a concentration of 1000 ppb was injected uniformly along the hole, and time series concentrations were measured by a Fluorometer (Turner Design, San Jose, CA, USA). The rate of decrease in tracer concentration was used to calculate the horizontal groundwater velocity in the well by the procedure described in Reference [50]. Results are given in The Supplementary Materials.

Interference recovery test was conducted by analyzing the level recovery in wells 71-2 (unit B) and 71-1 (unit C) after stopping pumping in the nearby (within 50 m) well 211 (Figure 1). Test results were used to calculate the transmissivity and the storativity of the aquifer, using the solution of Reference [51] for confined aquifers.

Radon ($^{222}$Rn) was sampled in both seawater and groundwater, using pre-vacuumed 4.5 L glass bottles. Seawater was sampled close to the seafloor in coastal water up to 700 m offshore (maximum water depth of 5 m) and groundwater was sampled from wells (the 71 series) using a peristaltic pump. $^{222}$Rn activities were measured by alpha-scintillation Lucas cells, using a modified emanation technique. In this method, helium is circulated and bubbled through the sampling bottle, resulting in the extraction of $^{222}$Rn. The radon is trapped on activated charcoal (at −90 °C) and then transferred to the Lucas cell.

Numerical simulations were conducted with the 2D FEFLOW program, which is a finite element code, allowing the examination of both flow and salt transport. To deal with the specific process in a precise way, a very dense net of triangular elements (~100,000) was built (element size range from ~1 m to 10 m). The model configuration included seven layers, in accordance with the geological section of the southern Carmel coastal plain: five layers in the Quaternary section and two in the underlying Late Cretaceous (Figure 2b). The conceptual model and model calibration were based on field measurements, presented and discussed in Section 4, including seabed geology (provided by the CHIRP survey), the TDEM survey, EC profiles, hydrological field experiments, and groundwater levels. The length of the model section is 7 km, from the sea on its west to the Carmel Mountains on its eastern end.

## 4. Field Results

### 4.1. Water Level

Figure 4a presents sea level and groundwater levels at the southern (wells 71-1, 71-2, 71-3) and the northern sites (well 73, Figure 1). Groundwater table in the intermediate confined unit (B) at the southern site was 0–0.3 m asl, while in the phreatic unit (A) and the deeper confined unit (C) it was higher, 1.5 and 1.3 m asl, respectively. The level in well 71-2 (unit B) and well 71-1 (unit C) was affected by a nearby pumping well, which is reflected in level increase during pumping breaks (i.e., return to more natural conditions) by 80 and 20 cm, respectively. At the northern site, water level in Unit B was higher (~1 m asl). In Figure 4b, we compare sea level and groundwater levels by zeroing to their mean values. Tidal fluctuation was hardly observed in the phreatic unit (southern site) at 70 m from the sea (1 cm), while it was clearly evident in the confined units B and C (amplitudes of up to 13 and 9 cm, respectively, during spring tide, compared with 21 cm at the sea, Table 1). At the northern site, maximum amplitude in unit B at 70 m from the sea and during the same period was just 7 cm. Level fluctuation decreased with distance from shore (4.5 and 3 cm at 300 m and 500 m, respectively, at the 2 and 1 sites, Table 1).

**Table 1.** Maximum tidal amplitudes in different units and distance from shoreline during May 2014.

| Amplitude (cm) | Distance from Shoreline (m) | Aquifer Unit | Well Name |
|---|---|---|---|
| 21 | | | Sea |
| 0.5 | 70 | A—south | 71-3 |
| 13 | 70 | B—south | 71-2 |
| 9 | 70 | C—south | 71-1 |
| 7 | 70 | B—north | 73 |
| 4.5 | 300 | B | 2 |
| 3 | 500 | B—north | 1 |

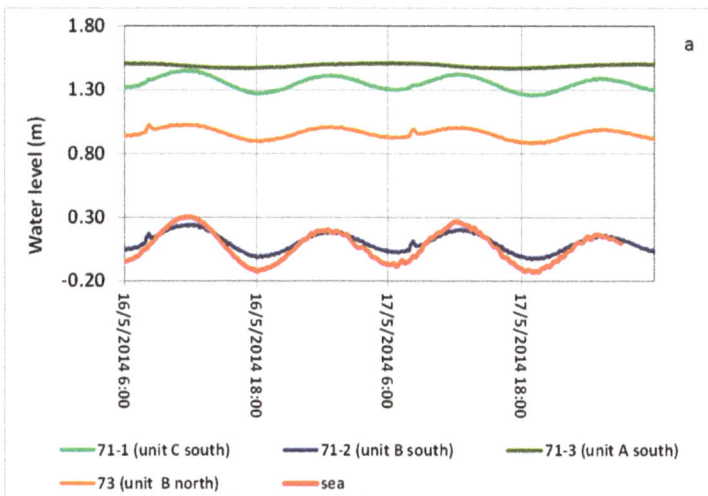

**Figure 4.** *Cont.*

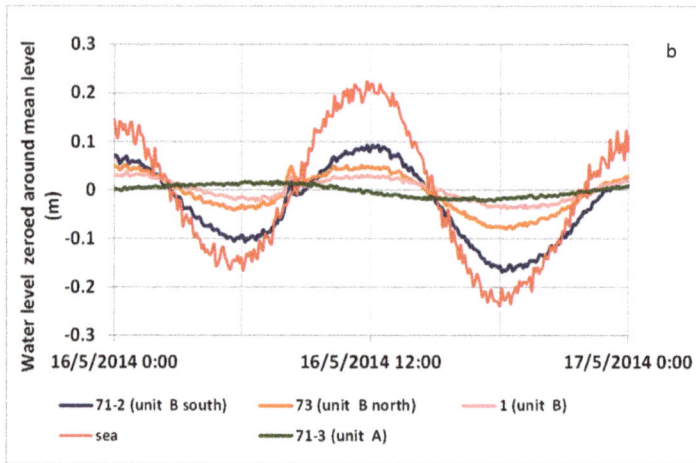

**Figure 4.** Detailed (5 min) water level time series in monitoring wells and the sea. (**a**) a 48 h level fluctuations; (**b**) 24 h superimposed levels of different wells and the sea, zeroed around their mean level. See well locations in Figure 1.

*4.2. EC Profiles*

Electrical conductivity profiles in units C, B, and A, at 70 m from shore, are presented in Figure 5. Units A and C show low and uniform EC values (5–6 mS/cm, similar to the natural salinity in this area) while unit B shows an interface composed of two steps. The EC changes from 5.5 mS/cm at a depth of 12 m to 40 mS/cm at a depth of 24 m.

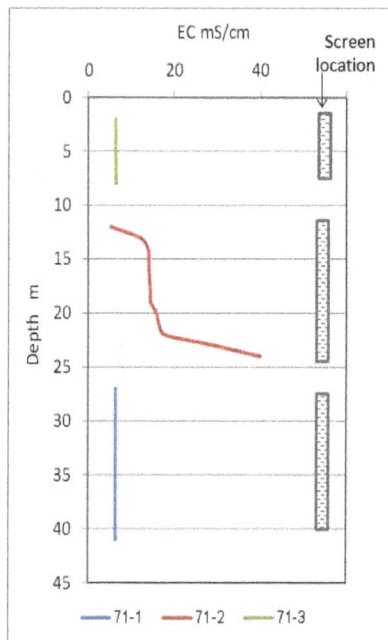

**Figure 5.** EC profiles in wells 71-3, 71-2, 71-1 penetrating the three Quaternary units (A, B and C, respectively), 70 m from the sea, at the southern site of Ma'agan Michael.

## 4.3. Electrical Resistivity—TDEM

The interpreted TDEM results from the southern site are in agreement with the EC borehole profiles. According to the TDEM, there is a large resistivity difference in unit B between the southern and the northern sites (Figure 6 and Figure S1 in the Supplementary Material), whereby relatively fresh groundwater is found in the northern site (resistivity of about 10 ohm-m), while brackish groundwater is exhibited in the southern site (resistivity of 3–5 ohm-m). The TDEM results also suggest that relatively fresh water occurs in unit C at both sites (Figure 6 and Figure S1 in the Supplementary Material).

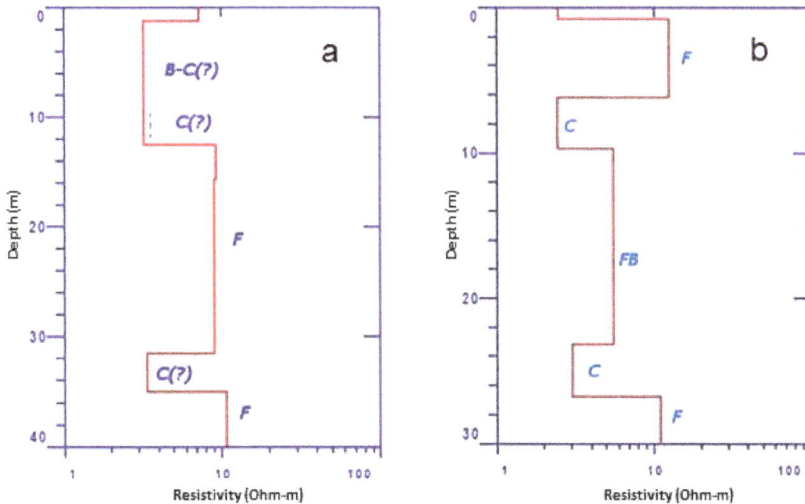

**Figure 6.** Representative TDEM (Time Domain Electro Magnetic) profiles at the northern site (**a**) and at the southern site (**b**). In the north, high resistivity is observed for both unit B and C (10 Ohm-m), while in the south the middle aquifer (unit B) has low resistivity (5 Ohm-m). The letters along the profile are the interpretation of the resistivity values: F—Fresh, B—Brackish, FB—fresh–brackish, C—Clay unit. Additional TDEM profiles are included in Figure S1 in the Supplementary Material.

## 4.4. Seismic Reflection

A CHIRP survey was conducted along eight parallel E–W offshore transects, 2 km each (Figure 7). The interpretations of the seismic data allowed the identification of the top of the upper clay and of the intermediate Kurkar unit (B). CHIRP results also show areas, where the clay layer is missing and the Kurkar rock is exposed at the seabed (e.g., Figure 7). The results of the survey suggest that there is a major difference between the southern and the northern sites with regards to the continuation of the upper clay layer (Figure 8). While at the southern site the clay was eroded around 100 m from shore line, at the northern site the clay layer is mostly continuous, and if missing, it occurs in limited areas relatively far from shore (>700 m). A sketch of the geological section, which expresses the difference between the southern and the northern site, is presented in Figure 9. The CHIRP survey also identified a "hot spots" (weak acoustic reflection in the high frequently CHIRP channel) along Sections 2 and 3, which was interpreted as groundwater discharge at one of the unit B exposure sites (Figure S8 in the Supplementary Material).

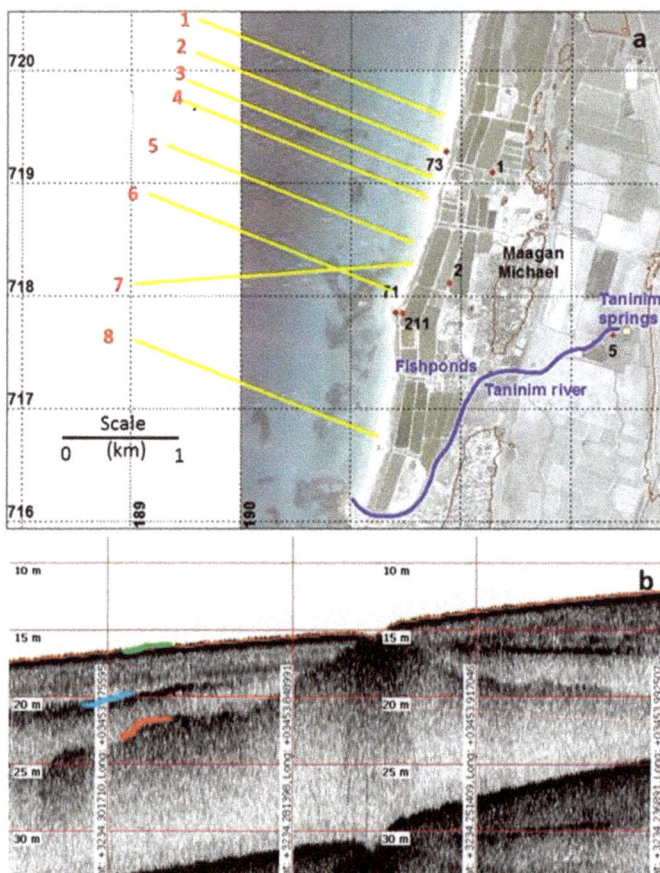

**Figure 7.** Location map of the CHIRP survey lines (**a**) and CHIRP seismic interpretation along part of Section 2 (**b**). The top calcareous sandstone ('Kurkar') layer is shown by red line, the top clay layer by blue line and the seabed by green line.

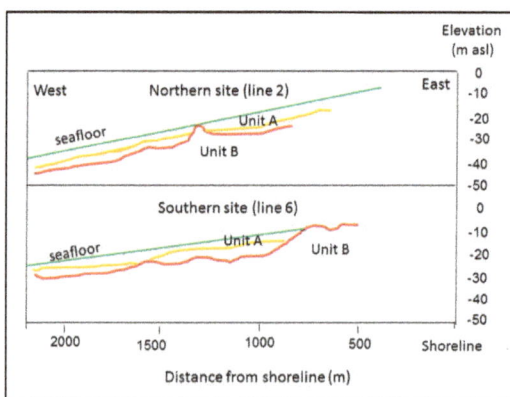

**Figure 8.** Interpretation of CHIRP lines 6 and 2 (location in Figure 8). The top Kurkar layer is shown by red line, the top clay layer by orange line and the seabed by green line.

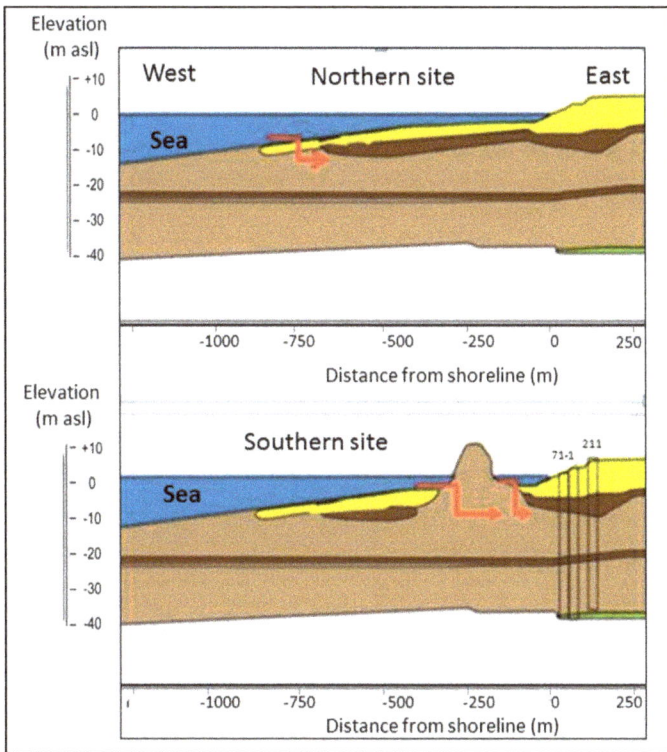

**Figure 9.** Schematic geological section in the northern and southern sites. In the southern site, the clay is eroded near the shore (100 m), therefore, the Kurkar unit is in direct contact with the sea closer to shore.

*4.5. Radon*

The average Radon activity in water sampled onshore from the confined Kurkar units B and C (wells 71-2 and 71-1, Table 2) was 735 dpm/L, quite similar to the findings at Dor Bay, 5 km to the north [5]. Most seawater samples (10–800 m from shore, all taken close to seabed) had much lower activities (0–2 dpm/L). One exception of 10.9 dpm/L was found at 5 m depth, above a seabed Kurkar exposure, about 700 m offshore at the northern site (Figure S9 in the Supplementary Material).

**Table 2.** Radon activity in groundwater and fish pond water (dpm/L) from Aug 2010 and May 2011.

| Rn | Source | | Sample |
|---|---|---|---|
| **dpm/L** | | | |
| 841 | Calcareous sand stone | Units B + C | 71-1 |
| 628 | Calcareous sand stone | Units B + C | 71-2 |
| 5 | Fishpond | Pond | Fish pond |

*4.6. Hydrological Tests*

Table 3 summarizes the hydraulic parameters, obtained from the field experiments, which are later used as the initial hydraulic parameter for the model calibration (Section 5). The slug and the interference recovery tests yielded similar hydraulic conductivity values, 45 and 68 m/day in unit B and 66 and 109 m/day in unit C, respectively. Since the slug is a local test, which is influenced by the well structure, we prefer to use the values obtained from the interference test, whenever available. The detailed results of the field experiments are given in Figures S2–S7 and Tables S1–S3 in the Supplementary Material.

**Table 3.** Hydraulic Transmissivity, conductivity, and Storativity, determined by field tests (slug, point dilution, and the interference recovery tests).

| Unit A (South) | Unit B (South) | Unit C (South) | Unit B (North) | |
|---|---|---|---|---|
| 195 | 750 | 1200 | 780 | T (m²/day) |
| 30 | 68 | 109 | 71 | K (m/day) |
| | 0.0001 | 0.0016 | | S |

Point dilution results from well 71-3 suggest that flow rate (v) in unit A was 5.9 m/day. Considering the hydraulic gradient near the shore (j = 1.5 m/70 m), and assuming porosity (n) of 0.15 for the sandy matrix, the hydraulic conductivity for the phreatic sand unit is 42 m/day (k = nv/j). This value is somewhat higher than the value of 13.2 m/day obtained from the slug test conducted in the same well. For the model calibration, we choose an intermediate value of 30 m/day.

## 5. Numerical Modeling

### 5.1. Geological Configuration

The eastern boundary was set at ~2 km east of the Taninim springs, where Late Cretaceous dolomite rocks are exposed at the foot of Mt. Carmel. This allowed the examination of the effect of pumping from the coastal Quaternary aquifer on the Late Cretaceous aquifer. The western boundary was set 3 km offshore, based on the CHIRP results, which showed that the onshore hydrogeological configuration continues beneath the seabed. Based on the offshore differences between the northern and the southern sites (see CHIRP results above), we conducted simulations with two different hydrogeological configurations. In the first, the upper clay unit extended 800 m offshore, while in the second, simulating the southern site, it reached just 100 m offshore (Figure 9).

### 5.2. Hydraulic Properties

The hydraulic conductivity of the Late Cretaceous dolomites was assumed to be 200 m/day [52], while for the Late Cretaceous chalks, underlying the Quaternary aquifer, a value of 1 m/day was assigned. Hydraulic parameters in the coastal aquifer were estimated according to the pumping results and the slug tests, conducted during this study (Table 2). After calibration, the hydraulic conductivity was taken to be somewhat higher than field estimates, 100 and 160 m/day for units B and C, compared with 66 and 100 m/day, respectively. On the eastern part, where no division to sub-aquifers was observed, a value of 100 m/day was assumed for the whole Quaternary aquifer. Conductivity of the phreatic sand unit (A) was assumed 30 m/day (between the 42 m/day of the point dilution results and 13 m/day of the slug test). The specific storage was assumed $10^{-4}$ m$^{-1}$ for both units, porosity was taken as 0.1, and dispersion as 5 m/day, with anisotropy of 10:1 (longitudinal to transversal ratio).

### 5.3. Boundary and Initial Conditions

We used two extreme scenarios for water level at the eastern boundary: level of +5 m, as was used to be in the more pristine conditions [37], and level of +3 m, as is the current case due to over-pumping from the Late Cretaceous Yarkon–Taninim carbonate Aquifer. The head at each offshore point was calculated as water depth multiplied by the seawater density. The salinity at the western boundary was taken as 39,000 mg/L TDS (Total Dissolved Solutes) and the specific gravity was 1.027, similar to that of the coastal Mediterranean Sea. Based on estimates of the Hydrological Survey of Israel, we used infiltration coefficient of 35% (average of 200 mm/year) for the western 0.7 km of the coastal plain, which is covered by sand and some clay, while a coefficient of 18% was assumed for the clayey soils of the eastern part. Total ponds losses were assumed as infiltration of 13 mm/day (about 3 m$^3$year$^{-1}$ for 1 km of coast), following [40].

The simulation was performed with the above conditions until steady state was reached. These steady-state conditions were used as initial conditions for the transient modeling after pumping began.

### 5.4. Pumping Regime

Withdrawal from the coastal aquifer in the studied area during 2005–2015 was 18–20 m$^3$year$^{-1}$. According to chemical balances [12], the fish ponds' contribution to the Kurkar aquifer (units B) is 2–3 × 10$^6$ m$^3$year$^{-1}$. This value was subtracted from the total pumping instead of adding artificial recharge from the pond, and, thus, the effective pumping was assumed ~15 × 10$^6$ m$^3$year$^{-1}$. Since the length of the coast in the studied area is about 3 km, the pumping in the model was taken as 5 × 10$^6$ m$^3$year$^{-1}$ per 1 km coast. Most pumping wells are perforated along the whole Pleistocene section and pump from both confined units (B and C). Therefore, the model assumes that of the existing four pumping wells, two pump from unit B and the other two from unit C (Figure 2b, red points).

### 5.5. Calibration of the Model

The calibration was done by comparing the dynamic results of the simulation to field measurements of water level and salinity in the different aquifer units at the southern site, near the sea. This includes: (1) water level in units C and B (1.3 and 0.3 m, respectively); (2) water salinity in unit B at the site (occurrence of FSI), and (3) low salinity in unit C. Because the pristine level conditions in the different units are not known, we calibrated the model only for the dynamic condition (pumping of 5 × 10$^6$ m$^3$year$^{-1}$ per 1 km coast). The simulation was conducted for the maximum and minimum heads at the eastern boundary (+5 and +3 m). Both scenarios were simulated until the flow field stabilized.

## 6. Simulation Results

### 6.1. Seawater Intrusion

The fresh–saline water interface in unit B is located where the confining clay layers breach the seabed, which is further off-shore at the northern site than in the southern site (Figure 10a,b). The location of the interface does not change much after pumping in the case of an eastern boundary head of +5 m (not shown in Figure 10), while in the case of +3 m head, pumping results in significant inland shift of the interface in unit B (Figure 10c). The location of the interface after 1000 days of pumping is shown in Figure 10c,d, and the change in salinity in unit B at a distance of 70 m from the shore (in observation point 2, Figure 10) is shown in Figure 11. This shift occurs both at the southern and at the northern site, but while at the southern site the interface reached the pumping wells, no salinization occurred at the northern site during the first few years (Figure 11). At both sites, penetration of seawater in unit C was very limited.

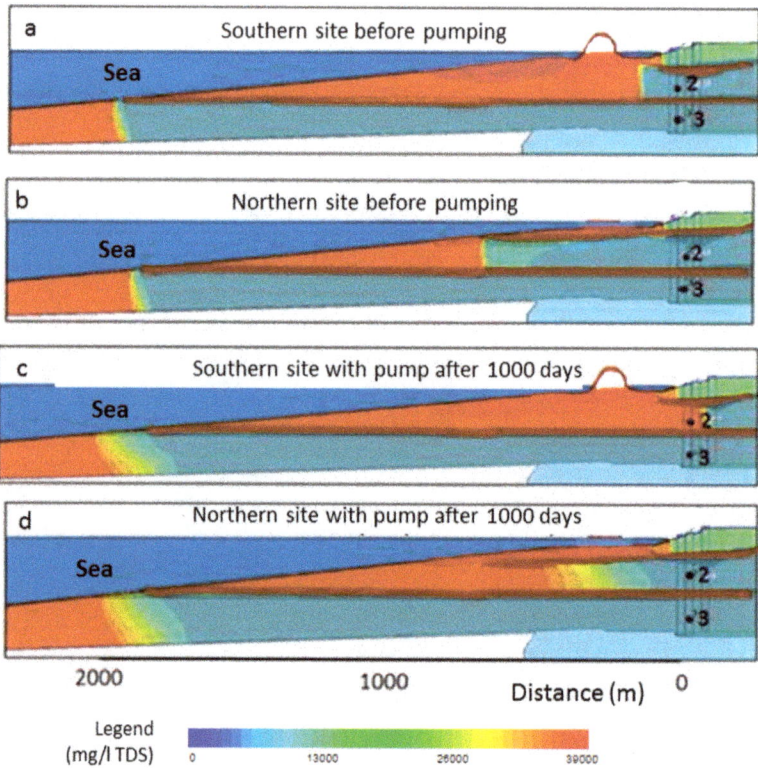

**Figure 10.** The location of the fresh–saline water interfaces in the different sub-aquifers: before pumping in the southern site (**a**); and in the northern site (**b**); after 1000 days of pumping with eastern boundary water level of +3 m, in the southern site (**c**) and the northern site (**d**). "2" and "3" denote units B and C.

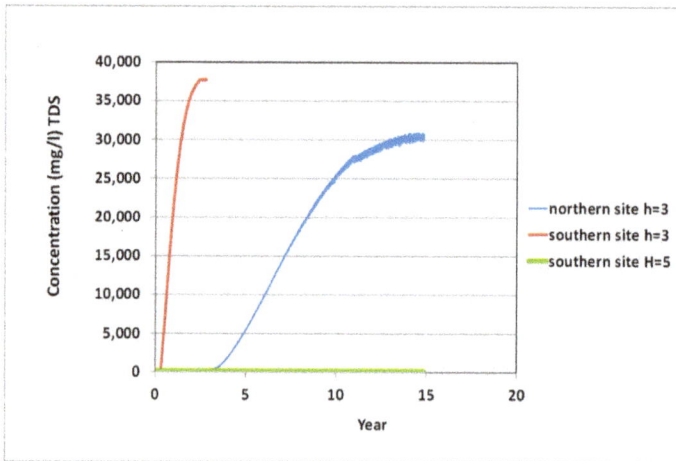

**Figure 11.** The simulated salinity in unit B at a distance of 70 m from the shore in the northern and southern sites (point 2 in Figure 10), for different boundary conditions (level of +5 and +3 m at the eastern model border).

### 6.2. Water Balance and Levels in Units B and C

The calculated water levels in observation boreholes at the pristine state and after pumping are given in Table 4. To maintain the average water level of +1.3 m in unit C at 70 m from shore (as observed in the field, Figure 4a), a significant amount of water must enter from the base of the coastal aquifer, namely from the Late Cretaceous aquifer. According to the results of the model, the increase in pumping from the coastal aquifer causes a ~0.4 m decrease in Late Cretaceous water level around the Taninim springs (borehole 5, Figure 1). The calculated water volumes entering from the Late Cretaceous aquifer to the coastal aquifer are shown in Table 5. In the natural state, when the water level in the eastern boundary was ~+5 m, this flux is estimated at $2.2 \times 10^6$ m$^3$year$^{-1}$ per 1 km of coast, which is almost the natural flow to the sea. After the initiation of pumping from the coastal aquifer ($5 \times 10^6$ m$^3$year$^{-1}$ per 1 km coast), flux from the east increased to $5.4 \times 10^6$ m$^3$year$^{-1}$ per 1 km, and the discharge to the sea from units B and C almost ceased (Table 6). With an eastern boundary water level of +3 m (over-exploitation of the Late Cretaceous Yarkon–Taninim aquifer), the pristine water flow into the coastal aquifer was smaller ($1.7 \times 10^6$ m$^3$year$^{-1}$ per 1 km coast), and after pumping commenced, flux from the east increased to 4.6 m$^3$year$^{-1}$ per 1 km (Table 5). Under these conditions, discharge from the confined units to the sea is negative (unit B: −0.7 and unit C: −0.2 $\times 10^6$ m$^3$year$^{-1}$, Table 6), and seawater intrudes as to keep the water balance in these units.

**Table 4.** Simulation-calculated water levels in observation boreholes, before and after pumping in the southern and northern sites under different boundary conditions. The measured levels in wells 71-1, 71-2, which represent the level after pumping, are presented in Figure 4a.

| Site | Level in the Eastern Border | Pump Condition | Well T5 | Well 71-2 | Well 71-1 |
|---|---|---|---|---|---|
| | | | Late Cretaceous | Unit B | Unit C |
| Southern site | H = +5 m | No pump | 4.6 | 1.2 | 2.8 |
| | | With pump | 4.2 | 0.6 | 1.4 |
| Southern site | H = +3 m | No pump | 2.8 | 0.8 | 1.9 |
| | | With pump | 2.4 | 0.3 | 0.8 |
| Northern site | H = +3 m | No pump | 2.8 | 1.8 | 2.2 |
| | | With pump | 2.3 | 0.3 | 0.8 |

**Table 5.** Calculated water input from the Late Cretaceous aquifer to the coastal aquifer in the pristine case and after pumping of $5 \times 10^6$ m$^3$year$^{-1}$ per 1 km coast from the coastal aquifer, with eastern boundary level of +5 m and + 3 m.

| East Border Head (m) | Unit | Q ($10^6$ m$^3$year$^{-1}$ per 1 km) | Q ($10^6$ m$^3$year$^{-1}$ per 1 km) |
|---|---|---|---|
| | | without Pumping | with Pumping of $5 \times 10^6$ m$^3$year$^{-1}$ per 1 km |
| H = +5 | A | 0.4 | 0.3 |
| | B | 1.2 | 2.6 |
| | C | 0.6 | 2.5 |
| | SUM | 2.2 | 5.4 |
| H = +3 | A | 0.2 | 0.2 |
| | B | 1.2 | 2.1 |
| | C | 0.3 | 2.3 |
| | SUM | 1.7 | 4.6 |

**Table 6.** Calculated flux to the sea with and without pumping from the coastal aquifer. In brackets, is the flux from unit A affected by ponds losses. The natural discharge to the sea is about 2.2 m$^3$year$^{-1}$ per 1 km.

| Eastern Border Head (m) | Unit | Q ($10^6$ m$^3$year$^{-1}$ per 1 km) | Q ($10^6$ m$^3$year$^{-1}$ per 1 km) |
|---|---|---|---|
| | | without Pumping | with Pumping of $5 \times 10^6$ m$^3$year$^{-1}$ per 1 km |
| | A | 0.4 (1.8) | 0.3 (1.8) |
| | B | 1.2 | −0.1 |
| H = +5 | C | 0.6 | 0 |
| | SUM | 2.2 | 0.2 |
| | A | 0.2 (1.8) | 0.2 (1.8) |
| | B | 1.2 | −0.7 |
| H = +3 | C | 0.3 | −0.2 |
| | SUM | 1.7 | −0.7 |

*6.3. Flow in Unit A*

The model results show that the estimated losses of $3 \times 10^6$ m$^3$year$^{-1}$ per 1 km coast from the ponds are divided between infiltration into unit A, which then flows directly to the sea, and unit B (about 1.8 and $1.2 \times 10^6$ m$^3$year$^{-1}$, respectively). The simulated water level in unit A varies between 5.5 m upstream and 1.5–2 m near the coast, which fits the measured levels (+5.5 m in well 2a and +1.5 m in well 71-3, 300 and 70 m from the coast, respectively, Figures 1 and 4a). The estimated flux from unit A between well 71-3 and the sea, using Darcy equation (J = 1.5 m/70 m, T = 195 m$^2$/day, Table 3) is $1.5 \times 10^6$ m$^3$year$^{-1}$ per 1 km coast.

*6.4. Tidal Effects*

The difference in sub-sea extension of the confined layer B between the southern and northern site results in different tidal amplitudes and water level in the observation wells 70 m on shore. The simulations suggest larger tidal amplitudes and lower level in unit B at the southern site, compared with the northern one (Figure 12), in agreement with our observations (wells 71-2 and 73, Table 1, Figure 12).

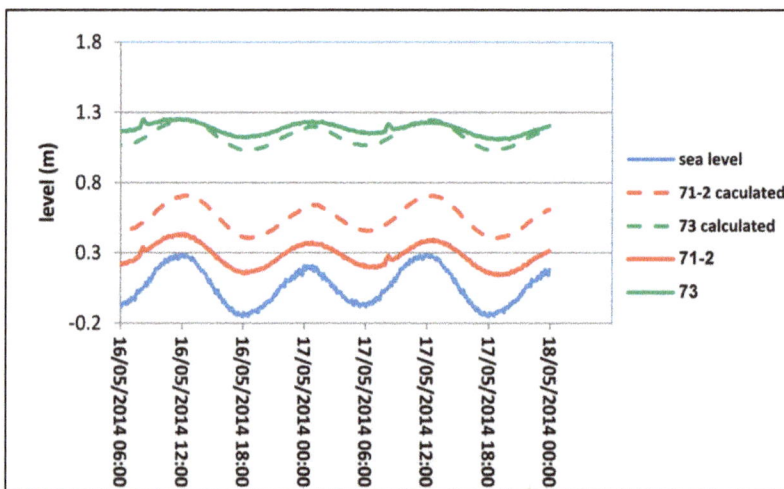

**Figure 12.** Measured and calculated water levels in observation wells in unit B, located 70 m from shore in the northern site (well 73) and in the southern site (well 71-2).

## 7. Discussion

### 7.1. Offshore Extension of the Confining Layer and Seawater Intrusion

The above simulations, as well as field results, highlight the importance of the boundary hydrological conditions on seawater intrusion in a multi-layered coastal aquifer. The difference in the seaside boundary conditions is expressed in the numerical simulations in different locations of the interface zone. While the interface at the southern site reaches the shoreline within a few years after pumping began, the interface at the northern site, where the clay layer extends farther seawards, is still located under the sea. These results are consistent with those of the TDEM, which showed relatively fresh groundwater in unit B of the northern site, while a clear interface at the southern site. The simulations and the field results show that the confined clay layer continuity in the sea has an important role in delaying seawater intrusion even when the aquifer is extensively exploited. The difference in the sub-sea extension of the upper confining clay layer also results in a difference in the groundwater tidal amplitudes, which are larger at the southern site (Table 1, Figure 4b), where the clay is missing closer to shore, than at the northern site, where the clay is mostly continuous to at least 700 m from shore. This difference in tidal amplitudes is also reflected in the simulations (Figure 12), although the measured amplitudes at borehole T-73 are somewhat smaller than the simulated ones. This difference could be eliminated by assigning larger storativity to the aquifer. However, this will cause a significant lag in the arrival of a tidal signal in the aquifer, in disagreement with our observations (Figure 13). Alternatively, we suggest that the relatively small amplitudes are due to the leaky nature of the thin confining layer, which would tend to damp the amplitude in the inland aquifer [26]. The numerical model also highlighted the importance of the land side (eastern) boundary condition, in the Late Cretaceous aquifer. Using high boundary heads (natural condition, before pumping started), seawater intrusion does not occur even under high exploitation, while with lower boundary heads, the aquifer is more liable to seawater intrusion.

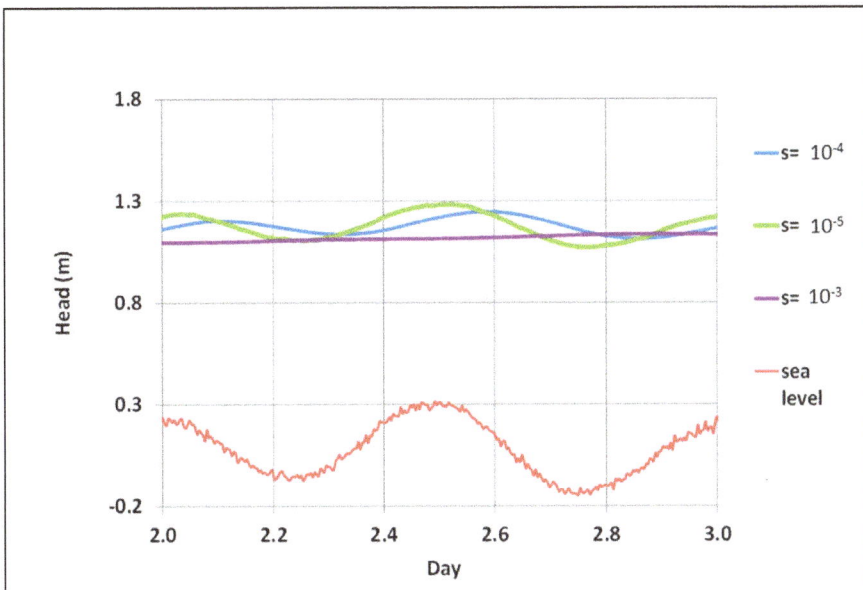

**Figure 13.** Transient heads at observation point 2 in the northern site (indicated in Figure 10, representing borehole 73) for various values of storativity/specific staorge of the confined unit B. Using high storativity, the aquifer tidal signal lags significantly after the sea, which is in disagreement with our observations.

*7.2. The Impact of Pumping on the Adjacent Aquifer*

The numerical model was used to quantify the flow regime into the multi-layered aquifer system. Previous works [12,38–40] showed hydrogeological and geochemical evidence for contribution from the Late Cretaceous aquifer to the coastal aquifer, whose extent was not clear. Our modeling examined more quantitively the suggestion that the Late Cretaceous aquifer is the main source of water (about 85%, [12]) to the coastal Quaternary aquifer (units B and C). The results of the simulations support this suggestion and show that increasing exploitation from the coastal aquifer causes an increase in the component of the water arriving from the adjacent Late Cretaceous aquifer. This increase in pumping also affects the discharge of the nearby Taninim springs. The model shows that to maintain the average water level of +1.3 m in the confined unit C at 70 m from shore under pumping conditions, a significant amount of water must enter from the base of the coastal aquifer, namely from the Late Cretaceous aquifer at the foot of Mt. Carmel, where the Quaternary rocks are in hydraulic contact with Late Cretaceous dolomites. This is possible due to the high hydraulic conductivity in both Pleistocene and the Late Cretaceous aquifer (100–200 m/day).

*7.3. Discharge to the Sea*

According to water budgets, water losses from the fishponds to the aquifer are about $9 \times 10^6$ m$^3$year$^{-1}$ [40]. The connection of the ponds with units B and C is limited due to the existence of a clay layer under the ponds area [12], which limits infiltration to 2–3 $\times 10^6$ m$^3$year$^{-1}$. This implies that most pond losses flow through unit A to the sea. The simulation results show that although unit A is of very small thickness (up to 6 m), it transfers $1.8 \times 10^6$ m$^3$year$^{-1}$ per 1 km coast, while keeping the level at +5.5 m and +1.5 m at 300 m and 70 m from the sea, respectively. These are, indeed, also the levels that were measured in the field. Considering fishpond length of 3 km, the flow from unit A to the sea is 5.4 m$^3$year$^{-1}$. According to simulation results, the flow from units B and C to the sea after the development of pumping from the coastal aquifer is relatively small and changes according to the eastern boundary condition between almost zero to negative values.

Radon activity in the calcareous sandstone (Kurkar) units was around 450 dpm/L, similar to the activities measured in groundwater from Dor bay, ca. 5 km to the north, where Reference [5] showed that the SGD radon source is the fresh groundwater and used it to calculate the fresh water discharge to the bay. Areas where the confided clay layer is eroded and the Kurkar is exposed on the seabed are suitable for Submarine Groundwater Discharge. Indeed, a radon hot spot (10.9 dpm/L) was found 500 m offshore at the northern site, where SGD was inferred also by the CHIRP results. Based on the radon–salinity mixing line between seawater and the Kurkar water source, the seawater at this point contains at least 2.5% freshwater derived from the Kurkar unit.

Since groundwater in unit A is highly polluted in several places [12], the effect of this SGD on the quality of the shallow sea could be significant.

**8. Summary and Conclusions**

This research combines field measurement and numerical simulation to examine the interrelations between a multi-layered coastal aquifer, a surface reservoir (fish ponds) and the sea. The measurement includes offshore geophysical surveys (CHIRP) and on land TDEM, together with hydrological tests and water level measurements. The main conclusions of this research are:

(1)    The location of the hydraulic connection between a confined sub-aquifer and the sea (the distance from the shoreline) is shown to be a very important factor in the determination of the extent of seawater intrusion.
(2)    The location of the aquifer-sea connection also affects the tidal fluctuations in the monitoring boreholes.
(3)    Submarine CHIRP mapping indicates that the continuity of the confining clay varies on short spatial scales (2 km).

(4)  In the Numerical model, this difference in clay continuity resulted in different extent of seawater intrusion in the shallow confined layer.

(5)  Numerical simulations indicate that water level, as well as seawater intrusion in coastal aquifers, could be highly affected by its connection with an underlying aquifer.

We show that the exact location of the connection between the confined aquifer unit and the sea plays a significant role on the sensitivity of an aquifer unit to seawater intrusion. These geophysical methods need to be used to determine this location in the different part of the coastal aquifer when deciding on the exact pumping regime of the aquifer. Another practical way to estimate this location is by the tidal amplitude in an observation well near the seashore. We, therefore, suggest that these methods be used as managerial tools in the vicinity of the sea to avoid large seawater intrusion.

**Supplementary Materials:** Supplementary Materials are available online at http://www.mdpi.com/2073-4441/10/10/1426/s1.

**Author Contributions:** A.T. did the field work, conduct the modeling and wrote the manuscript, Y.W. took part in writing and in the field study, S.W. was responsible for the simulations, M.G. did the TDEM study and Y.Y. took part in the modeling and the writing of the paper.

**Funding:** This project was partly funded by a Water Authority of Israel grant 4500445470 and by a USAID MERC grant M29-073.

**Acknowledgments:** We are grateful to people from the Geological Survey of Israel for their support: to Halel Lutzki for the field tests, to Iyad Swaed for his help with the new wells, to Haim Hemo for the EC profiles, to Nili Almog and Bat-Sheva Cohem for their help in drawing the figures. Thanks to Yehuda Shalem and the Radon laboratory team in the HU for the help in the Radon sampling and analyzing. We are grateful to Eldad Levi from the Geophysical Institute for the support in the TDEM survey and to Aric Golan from the IOLR to the support with the CHIRP survey and the seismic interpretation. Thanks to Einat Magal for the help with the "Point dilution test" and to Yossi Berchman from Maagan Michael for their cooperation.

**Conflicts of Interest:** The authors declare no conflicts of interest.

## References

1.  Bear, J.; Cheng, A.H.D.; Sorek, S.; Ouazar, D.; Herrera, I. (Eds.) *Seawater Intrusion in Coastal Aquifers: Concepts, Methods, and Practices*; Kluwer Academic Publishers: Dordrecht, The Netherlands; Boston, MA, USA; London, UK, 1999.

2.  Post, V. Fresh and saline groundwater interaction in coastal aquifers: Is our technology ready for the problems ahead? *Hydrogeol. J.* **2005**, *13*, 120–123. [CrossRef]

3.  Winter, T.C. Numerical simulation of steady state three-dimensional groundwater flow. *Water Resour. Res.* **1978**, *14*, 245–254. [CrossRef]

4.  Moore, W.S. Large groundwater inputs to coastal waters revealed by 226Ra enrichments. *Nature* **1996**, *380*, 612–614. [CrossRef]

5.  Weinstein, Y.; Burnett, W.C.; Swarzenski, P.W.; Shalem, Y.; Yechieli, Y.; Herut, B. The role of coastal aquifer heterogeneity in determining fresh groundwater discharge and seawater recycling: An example from the Carmel coast, Israel. *J. Geophys. Res.* **2007**, *112*, C12016. [CrossRef]

6.  Moore, W.S. The subterranean estuary: A reaction zone of ground water and sea water. *Mar. Chem.* **1999**, *65*, 111–125. [CrossRef]

7.  Woessner, W.W. Stream and fluvial plain ground water interactions: Rescaling hydrogeologic thought. *Ground Water* **2000**, *38*, 423–429. [CrossRef]

8.  Linderfelt, W.R.; Turner, J.V. Interaction between shallow groundwater, saline surface water and nutrient discharge in a seasonal estuary: The Swan–Canning system. *Hydrol. Process.* **2001**, *15*, 2631–2653. [CrossRef]

9.  Trefry, M.G.; Svensson, T.J.A.; Davis, G.B. Hypoaigic influences on groundwater flux to a seasonally saline river. *J. Hydrol.* **2007**, *335*, 330–353. [CrossRef]

10.  Herbeck, S.L.; Unger, D.; Wu, Y.; Jennerjahn, C.T. Effluent, nutrient and organic matter export from shrimp and fish ponds causing eutrophication in coastal and back-reef waters of NE Hainan, tropical China. *Cont. Shelf Res.* **2013**, *57*, 92–104. [CrossRef]

11.  Burford, M.A.; Costanzo, S.D.; Dennison, W.C.; Jackson, C.J.; Jones, A.B.; McKinnon, A.D.; Preston, N.P.; Trott, L.A. A synthesis of dominant ecological processes in intensive shrimp ponds and adjacent coastal environments in NE Australia. *Mar. Pollut. Bull.* **2003**, *46*, 1456–1469. [CrossRef]

12.  Tal, A.; Weinstein, Y.; Yechieli, Y.; Borisover, M. The influence of fish ponds and salinization on groundwater quality in the multi-layer coastal aquifer system in Israel. *J. Hydrol.* **2017**, *551*, 768–783. [CrossRef]

13.  Werner, A.D.; Bakker, M.; Post, V.A.; Vandenbohede, A.; Lu, C.; Ataie-Ashtiani, B.; Simmons, C.T.; Barry, D.A. Seawater intrusion processes, investigation and management: Recent advances and future challenges. *Adv. Water Resour.* **2013**, *51*, 3–26. [CrossRef]

14.  Boufadel, M.C. A mechanistic study of nonlinear solute transport in a groundwater-surface water system under steady state and transient hydraulic conditions. *Water Resour. Res.* **2000**, *36*, 2549–2565. [CrossRef]

15.  Mercer, J.W.; Larson, S.P.; Faust, C.R. Simulation of salt-water interface motion. *Groundwater* **1980**, *18*, 374–385. [CrossRef]

16.  Polo, J.F.; Ramis, F. Simulation of salt water–fresh water motion. *Water Resour. Res.* **1993**, *19*, 61–68. [CrossRef]

17.  Essaid, H.I. A multilayered sharp interface model of coupled freshwater and saltwater flow in coastal systems: Model development and application. *Water Resour. Res.* **1990**, *26*, 1431–1454. [CrossRef]

18.  Huyakorn, P.S.; Andersen, P.F.; Mercer, J.W.; White, H.O. Saltwater intrusion in quifers: Development and testing of a three-dimensional finite element model. *Water Resour. Res.* **1987**, *23*, 293–312. [CrossRef]

19.  Das, A.; Datta, B. Simulation of density dependent 2-D seawater intrusion in coastal aquifers using nonlinear optimization algorithm. In Proceedings of the American Water Resource Association of Annual Summer Symposium on Water Resource and Environmental Emphasis on Hydrology and Cultural Insight in Pacific Rim, Herndon, VA, USA, 1995; American Water Association: Herndon, VA, USA, 1995; pp. 277–286.

20.  Putti, M.; Paniconi, C. Finite element modeling of saltwater intrusion problems. In *Advanced Methods for Groundwater Pollution Control*; International Centre for Mechanical Sciences; Springer: New York, NY, USA, 1995; Volume 364, pp. 65–84.

21.  Guo, W.; Langevin, C. User's guide to SEAWAT: A computer program for simulation of three-dimensional variable density groundwater flow. In *Techniques of Water-Resources Investigations Book 6*; US Geological Survey Intrusion of the Gaza Strip Aquifer—Palestine: Alicante, Spain, 2002; Volume 1, pp. 245–254.

22.  Yechieli, Y.; Shalev, E.; Wollman, S.; Kiro, Y.; Kafri, U. Response of the Mediterranean and Dead Sea coastal aquifers to sea level variations. *Water Resour. Res.* **2010**, *46*, W12550. [CrossRef]

23.  Van der Kamp, G. Tidal fluctuations in a confined aquifer extending under the sea. *Int. Geol. Cong.* **1972**, *24*, 101–106.

24.  Li, G.; Chen, C. Determining the length of confined aquifer roof extending under the sea by the tidal method. *J. Hydrol.* **1991**, *123*, 97–104.

25.  Li, G.; Chen, C. The determination of the boundary of confined aquifer extending under the sea by analysis of groundwater level fluctuations. *Earth Sci.* **1991**, *16*, 581–589.

26.  Li, H.; Jiao, J.J. Tide-induced groundwater fluctuation in a coastal leaky confined aquifer system extending under the sea. *Water Resour. Res.* **2001**, *37*, 1165–1171. [CrossRef]

27.  Burnett, W.C.; Kim, G.; Lane-Smith, D. A continuous monitor for assessment of $^{222}$Rn in the coastal ocean. *J. Radioanal. Nucl. Chem.* **2001**, *249*, 167–172.

28.  Conant, B., Jr. Delineating and quantifying ground water discharge zones using streambed temperatures. *Ground Water* **2004**, *42*, 243–257. [CrossRef] [PubMed]

29.  Mulligan, A.E.; Charette, M.A. Intercomparison of submarine groundwater discharge estimates from a sandy unconfined aquifer. *J. Hydrol.* **2006**, *327*, 411–425. [CrossRef]

30.  Li, L.; Barry, D.A.; Stagnitti, F.; Parlange, J.-Y. Submarine groundwater discharge and associated chemical input to a coastal sea. *Water Resour. Res.* **1999**, *35*, 3253–3259. [CrossRef]

31.  Li, H.; Jiao, J.J. Tide-induced seawater-groundwater circulation in a multi-layered coastal leaky aquifer system. *J. Hydrol.* **2003**, *274*, 211–224. [CrossRef]

32.  Swarzenski, P.W.; Burnett, W.C.; Greenwood, W.J.; Herut, B.; Peterson, R.; Dimova, N.; Shalem, Y.; Yechieli, Y.; Weinstein, Y. Combined time-series resistivity and geochemical tracer techniques to examine submarine groundwater discharge at Dor Beach Israel. *Geophys. Res. Lett.* **2006**, *33*, L24405. [CrossRef]

33.  Burnett, W.C.; Dulaiova, H. Estimating the dynamics of groundwater input into the coastal zone via continuous radon-222 measurements. *J. Environ. Radioact.* **2003**, *69*, 21–35. [CrossRef]

34. Weinstein, Y.; Shalem, Y.; Burnett, W.C.; Swarzenski, P.W.; Herut, B. Temporal variability of Submarine Groundwater Discharge: Assessments via radon and seep meters, the southern Carmel Coast, Israel. In *A New Focus on Groundwater–Seawater Interactions*; Sanford, W., Ed.; IAHS Press: Wallingford, UK, 2007; Volume 312, pp. 125–133.

35. Bar Yosef, J. Investigation of Salinization Mechanism in the Taninim Springs. *Tahal Water Plan. Israel Tel Aviv Rep.* **1974**, 36. (In Hebrew)

36. Bar, Y. Hydrogeology and Geochemistry of Groundwater in the Benyamina-Um El Fahm area. Master's Thesis, The Hebrew University, Jerusalem, Israel, 1983. (In Hebrew with English Abstract).

37. Dafny, E. Groundwater Flow and Solute Transport within the Yarqon-Taninim Aquifer, Israel. Ph.D. Thesis, The Hebrew University of Jerusalem, Jerusalem, Israel, 2009. (In Hebrew with English Abstract).

38. Michelson, H.; Zeitoun, D.G. Desalinization in the Nahal Taninim Area. *Tahal Water Plan. Israel Tel Aviv Rep.* **1994**, 73. (In Hebrew)

39. Guttman, J. Defining Flow Systems and Groundwater Interactions in the Multi Aquifer System of the Carmel Coast Region. Ph.D. Thesis, Tel Aviv University, Tel Aviv, Israel, 1998. Mekorot Report No. 467.

40. Bar-Yosef, Y.; Michaeli, A. *Hydrological Situation in the Carmel Coastal Plain, Israel*; NRD Report; NRD—NR/507/06; NRD: Tel Aviv, Israel, 2006. (In Hebrew)

41. Michelson, C. *Carmel Coast Geology*; Tahal Report HG/70/025; Tahal: Tel Aviv, Israel, 1970. (In Hebrew)

42. Almagor, G. *The Mediterranean Coast of Israel*, 2nd ed.; Geological Survey of Israel: Jerusalem, Israel, 2005; 273p.

43. Tal, A.; Zilberbrand, M. *Water Sources for the Coastal Aquifer in Maagan Michael Area*; Inside Report; Hydrological Service of Israel: Jerusalem, Israel, 2010.

44. Gilad, D.; Malul, A.; Goldman, M.; Ronen, A. *Seawater Intrusion Mapping in the Coastal Plain, Israel, in the TDEM Method*; HIS Report 1990/489/132/843; HIS: Tel Aviv, Israel, 1990. (In Hebrew)

45. Goldman, M.; Gilad, D.; Ronen, A.; Melloul, A. Mapping of seawater intrusion into the coastal aquifer of Israel by the time domain electromagnetic method. *Geoexploration* **1991**, *28*, 153–174. [CrossRef]

46. Goldman, M.; Rabinovich, B.; Rabinovich, M.; Gilad, D.; Gev, I.; Schirav, M. Application of the integrated NMR-TDEM method in groundwater exploration in Israel. *J. Appl. Geophys.* **1994**, *31*, 27–52. [CrossRef]

47. Kafri, U.; Goldman, M.; Lang, B. Detection of subsurface brines, freshwater bodies and the interface configuration in between by the Time Domain electromagnetic (TDEM) method in the Dead Sea rift, Israel. *J. Environ. Geol.* **1997**, *31*, 42–49. [CrossRef]

48. Golan, A. *Shallow Geological Structure in the Sea Part of Maagan Michael Area*; IOLR report H50/2014; IOLR: Tel Aviv, Israel, 2014. (In Hebrew)

49. Lutski, H.; Shalev, E. *Slug Test for Hydraulic Conductivity and Technical Status Interpretation of Wells in the Coastal Plain, Israel*; GSI Report; GSI: Tel Aviv, Israel, 2010. (In Hebrew)

50. Ward, R.S.; Williams, A.T.; Barker, J.A.; Brewerton, L.J.; Gale, I.N. *Groundwater Tracer Test: A Review and Guidelines for Their Use in British Aquifers*; British Geological Survey: Nottingham, UK, 1998; p. 286.

51. Cooper, H.H.; Jacob, C.E. A generalized graphical method for evaluating formation constants and summarizing well field history. *Am. Geophys. Union Trans.* **1946**, *27*, 526–534. [CrossRef]

52. Guttman, Y.; Cronveter, L. *The Effect of Increasing Pumping on the Flow Regime in Maagan Michael—Zichron Yaakov Area*; Mekorot Report No. 1291; Mekorot: Tel Aviv, Israel, 2007. (In Hebrew)

*Article*

# Quantification of Submarine Groundwater Discharge in the Gaza Strip

**Ashraf M. Mushtaha [1,2,*] and Kristine Walraevens [2]**

[1]    Environmental Affairs and MIS Departments, Gaza Strip, Palestine
[2]    Laboratory for Applied Geology and Hydrogeology, Ghent University, Krijgslaan 281-S8,
       9000 Gent, Belgium; kristine.walraevens@ugent.be
*    Correspondence: amushtaha@gmail.com

Received: 29 September 2018; Accepted: 7 December 2018; Published: 10 December 2018

**Abstract:** Gaza Strip has suffered from seawater intrusion during the past three decades due to low rainfall and high abstraction from the groundwater resource. On a yearly basis, more than 170 million $m^3$ of groundwater is abstracted, while the long-term average recharge from rainfall is 24.4 million $m^3$/year. Submarine groundwater discharge (SGD) has never been studied in the Gaza Strip, due to lack of experience in this field, next to the ignorance of this subject due to the seawater intrusion process taking place. Continuous radon measurements were carried out in six sites along the Gaza Strip to quantify the SGD rate. The final result shows SGD to occur in all sampled sites. The range of SGD rates varies from 0.9 to 5.9 cm·day$^{-1}$. High values of SGD are found in the south (Rafah and Khan Younis governorates). The high values are probably related to the shallow unconfined aquifer, while the lowest values of SGD are found in the middle of Gaza Strip, and they are probably related to the Sabkha formation. In the north of Gaza Strip, SGD values are in the range of 1.0 to 2.0 cm·day$^{-1}$. Considering that SGD would occur with the measured rates in a strip of 100 m wide along the whole coast line, the results in a quantity of 38 million $m^3$ of groundwater being discharged yearly to the Mediterranean Sea along Gaza coast. Nutrient samples were taken along Gaza Strip coastline, and they were compared to the onshore wells, 600 m away from the Mediterranean Sea. The results show that SGD has higher $NO_3^- + NO_2^-$ than nutrient-poor seawater, and that it is close to the onshore results from the wells. This confirms that the source of SGD is groundwater, and not shallow seawater circulation. In a coastal strip of 100 m wide along the Gaza coast, a yearly discharge of over 400 tons of nitrate and 250 tons of ammonium occurs from groundwater to the Mediterranean Sea.

**Keywords:** SGD; SGD model; Radon; coastal aquifer; nutrient discharge; Gaza Strip

---

## 1. Introduction

Although Submarine groundwater discharge (SGD) may not play a significant role in the global water balance, there are reasons to believe that the geochemical cycles of some major and minor elements may be strongly influenced either by the direct discharge of fresh groundwater into the sea, or by chemical reactions that occur during the recirculation of seawater through a coastal aquifer system [1–3]. Groundwater contamination, being a wide-spread problem, SGD may bring pollution to coastal seawater [3,4]. SGD affects the nutrient balance of the sea near the coast, causing harmful algal blooms and changing the flora and fauna of coastal waters [5–10].

Several papers have emphasized the important role of recirculated seawater in the transport of solutes from aquifers to coastal water. Moore [2] proposed the term "subterranean estuary (STE)" for the aquifer zone, where recirculating seawater mixes with fresh groundwater, and where water–rock interaction affects the mobility of constituents, including nutrients, towards the sea [11]. Lebbe [12], as well as Werner and Lockington [13], have demonstrated that tidal fluctuations can induce saltwater recirculation in the upper aquifer, producing a saline water table in the coastal fringe.

Measuring SGD by traditional hydrogeological or water balance estimates may be off by several orders of magnitude, largely because of difficulties in constraining hydraulic conductivities. Yet, quantifying SGD is important, as concentrations of dissolved constituents in SGD are often greater than in surface waters, resulting in significant groundwater-derived solute contributions [10,14]. One potential means of evaluating groundwater pathways and fluxes into the coastal zone more accurately is through the use of natural tracers, where $^{222}$Rn is an excellent tracer [3,15–18]. The very large enrichment of $^{222}$Rn concentration in groundwaters over surface waters (usually 1000-fold or greater), and its unreactive nature and short half-life ($t_{1/2} = 3.83$ days) make $^{222}$Rn an excellent tracer to identify areas of significant groundwater discharge [3,19]. Continuous radon monitoring (RAD 7-device with RAD-AQUA) could provide reasonably high-resolution data on the radon concentration of coastal seawater at one location over time [20].

In the past three decades, researchers were focusing on the seawater intrusion phenomenon in Gaza Strip, while no one has tackled the reverse flow direction, to establish whether it may occur in the hydrogeological setting. This study focuses on the submarine groundwater discharge from different places along the Gaza coastline using continuous radon measurements with RAD7 and RAD-AQUA devices, beside nutrient analysis along the shoreline and onshore in monitoring wells.

## 2. Hydrogeological Background

Gaza Strip is a coastal area along the eastern Mediterranean Sea and lies at latitude 31°25′59″ N and longitude 34°22′34″ E. Gaza Strip forms a transition zone between the semi-humid coastal zone in the north, the semi-arid zone in the east, and the Sinai desert in the south. Gaza Strip has a surface area of 365 km$^2$ where more than 1.8 million inhabitants are living. Groundwater is the only water source for the population of Gaza [21].

The Kurkar Group stratigraphy of Gaza Strip near the coast subdivides the aquifer into four sub aquifers (sub-aquifers A, B1, B2 and C), separated by marine clay layers, whereas in the east, there is only one aquifer (Figure 1). Those sub-aquifers are semi-confined near the coast, except for the upper layer. The bottom of the aquifer is an impermeable layer known as the Saqiya Group (Miocene-Pliocene age), which is a very thick sequence of marls, marine shales, and clay stones [22]. The Saqiya Group pinches out at around 10 km to 15 km from the coast in the South of Gaza Strip, and the coastal aquifer rests directly on Eocene-age chalk. The saturated thickness of the Gaza coastal aquifer near the coast is around 180 m in the North West, while it is around 40 m at the North Eastern border, and in the South East, it is only few meters (5 m to 10 m; [23]). The Gaza Strip has more than 5000 wells distributed all over the area; 200 wells are used for the water supply system, and the rest are either agricultural or private wells [24].

As a result of aquifer exploitation and imbalance between recharge and abstraction in the past few decades, the groundwater level has dropped to more than 10 m below mean sea level in the southern part of Gaza Strip [21,24]. The yearly groundwater decline rate varies from north to south: in the north, it ranges between 10 to 30 cm, and in the south, it reaches 70 cm [25].

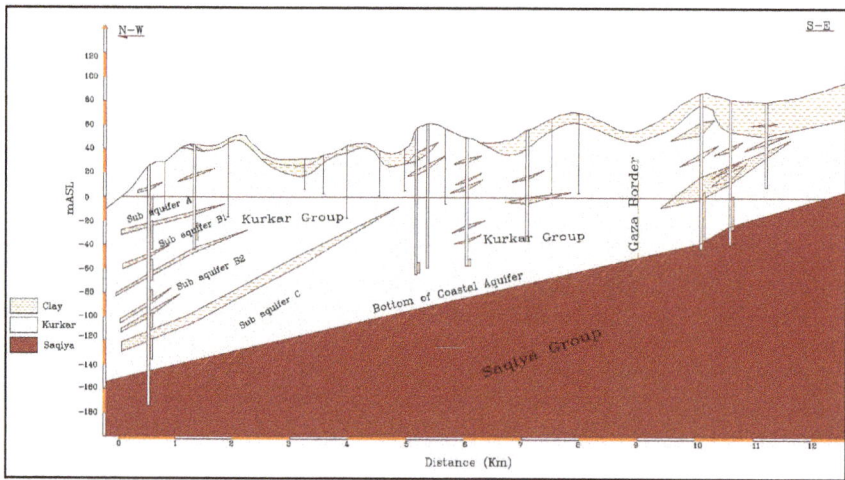

**Figure 1.** Typical cross-section of the Gaza Aquifer (after [26]).

## 3. Materials and Methods

### 3.1. Measurement of Nutrients in Groundwater

Coastal aquifers facilitate the interaction of nutrient-rich groundwater with nutrient-poor seawater. This is especially important in oligotrophic settings, such as the eastern Mediterranean, where the ($NO_3^-$ + $NO_2^-$) content of seawater is typically <100 µM [11,27]. Nutrient sampling of shallow groundwater was done to find indications of SGD, in order to select the best locations for radon measurements. A total of 232 water samples have been collected, 228 of them are groundwater samples taken along the Gaza coastline (51 samples in the north, 68 in the middle and 109 in the south), with a distance ranging from 10 to 20 m from the high-tide line, after digging for 1 to 1.2 m in the beach sand. Four samples were collected from shallow groundwater wells within 600 m from the coastline, which penetrate a few meters in the groundwater (in the south of the Gaza Strip). All collected samples were analyzed for nitrate ($NO_3^-$), nitrite ($NO_2^-$), ammonia ($NH_4^+$), phosphate ($PO_4^{3-}$), and salinity. Figure 2A shows the coastal samples' location (232 groundwater samples) and the grouping based on the sample location. Each group of coastal samples were averaged based on location: north, middle, and south of the Gaza Strip.

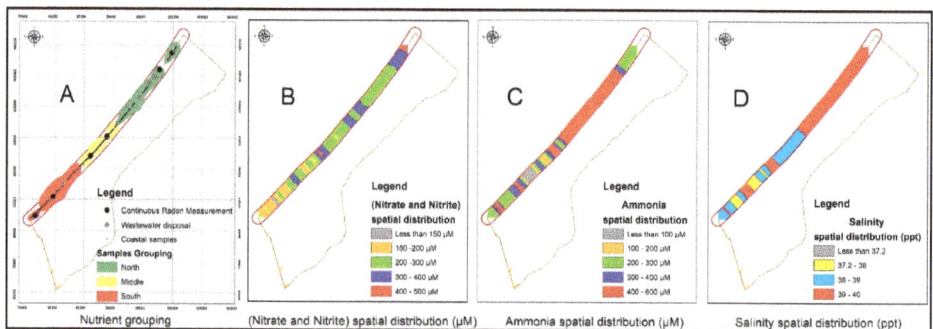

**Figure 2.** Nutrient sampling location and grouping, spatial distribution for $NO_3^-$ + $NO_2^-$, $NH_4^+$, and salinity (map grid according to the Palestinian Grid System 1923). (**A**) Samples grouping vs wastewater disposal location; (**B**) Spatial distribution of $NO_3^-$ + $NO_2^-$ along the coastline; (**C**) Spatial distribution of $NH_4^+$ along the coastline; (**D**) Spatial distribution of salinity along the coastline.

Samples locations were verified, to make sure that sewage outflow to the sea was avoided. Six locations were selected based on high nutrient concentrations to demonstrate SGD using RAD7 with RAD-AQUA for the continuous measurement of radon. Three maps of $NO_3^- + NO_2^-$, $NH_4^+$, and salinity were included to show the selection of the SGD locations (see Figure 2B–D).

*3.2. Measurement of Radon Concentrations in Groundwater*

Groundwater is enriched in radon compared to the surface water. While most of the radon ($^{222}$Rn) pore water is produced from particle-surface-bound radium ($^{226}$Ra), and the accumulation of the radium is likely regulated by the presence of manganese (hydr)oxides [28].

Radon has greater utility than radium as a SGD tracer when the discharge is fresh or a mixture of fresh and recirculated seawater, since any groundwater in the aquifer (independent of its salinity) is enriched in radon, due to its contact with the sediments and rocks. Radon-enriched fluid is transported to coastal waters, and a radon mass balance can be used to calculate total submarine groundwater discharge [28]. The main sources of uncertainty in the radon mass balance model are associated with:

(1)   quantifying $^{222}$Rn loss by evasion to the atmosphere [29],

(2)   quantifying $^{222}$Rn loss via mixing with offshore waters [30] and

(3)   characterizing the groundwater end-member radon activity that supplied the measured excess Rn in coastal waters [31–33].

Since radon is an inert gas, we may expect that the sediment and/or rock uranium/radium content is the main source of radon; it is not produced by chemical reactions or pore water chemistry [28]. The sole source of radon ($^{222}$Rn with a half time of 3.8 days) in groundwater is the radioactive decay of radium ($^{226}$Ra with a half time of 1600 years). Radium, the parent of radon, can be retained on grain surface coatings, bound in the mineral lattice of the aquifer matrix, and dissolved in pore water [2]. The short half-life of $^{222}$Rn makes it an excellent tracer to identify areas of significant groundwater discharge [34].

The automated radon system analyses $^{222}$Rn from a constant stream of water produced by a submersible pump and passed through an air–water exchanger, which distributes radon from the running flow of water to a closed air loop. The air stream is fed to the radon-in-air monitor. The radon concentration in the water can easily be calculated from the known temperature dependence of the radon distribution at equilibrium between the air and water phases [20].

Measurements of radon concentrations in the water column have been accomplished by standard oceanographic sampling and analysis techniques (radon emanation) for measurement of $^{222}$Rn taking the special care required for trace gas sampling [3,35,36]. Recently, "continuous" radon monitoring has been described, which can provide high-resolution data on the radon concentration of coastal seawater at one location over time [3,20].

The operation principle of the RAD AQUA device is the closed air loop in which the radon concentration reaches equilibrium during water supply flow. The radon concentration in the water compared to radon concentration in the air is governed by a temperature coefficient of 0.25 at room temperature [3], meaning that four times more radon is present in the air, compared to water. This has a favorable impact on measuring accuracy.

Six locations along the Gaza Strip coastline were measured for continuous radon, to assess the SGD (Figure 3): two locations in the north, two in the middle, and two in the south of the Gaza Strip.

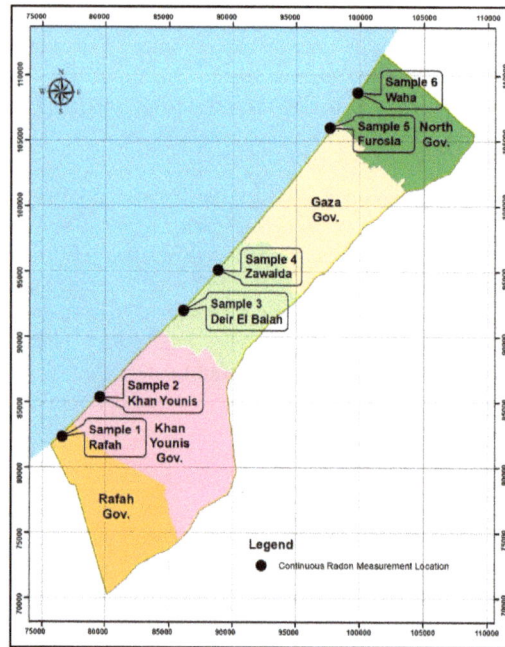

**Figure 3.** Continuous radon measurement location.

*3.3. Field Work*

To measure the submarine groundwater discharge using radon as a groundwater tracer, RAD7 and RAD-AQUA were used together to measure continuous radon in water in six locations in the Gaza Strip (Figure 3). A submersible pump was placed at the sea floor (at a depth ranging from 0.83 to 1.05 m), pumping water from the seafloor through a hose to the RAD-AQUA device for spraying, while the RAD7 measures the produced radon from the sprayed water. We initially intended to measure temporal changes of SGD, but due to the situation in the Gaza Strip, we were unable to do this. The time duration for sampling was different in each sampling location according to the field situation and obstacles found in the field. The duration ranged from 1.5 h to 2 days. In the middle part of Gaza Strip (Deir El Balah governorate), two short measurements were carried out, ranging from 1.5 to 3 h; due to some practical constraints, we were not able to spend more time for measuring at these sites. Table 1 shows the location and duration of each measurement.

**Table 1.** General information about the continuous radon measurement locations.

| Area | Site Name | Coordinates (Palestinian Grid System 1923) | | Continuous Radon Measurement Time Duration (Hours:Minutes) | Date of Measurement |
|---|---|---|---|---|---|
| | | X | Y | | |
| South | Rafah | 76,581.88 | 82,326.76 | 7:15 | 12 February 2014 |
| | Khan Younis | 79,585.25 | 85,385.75 | 12:30 | 11–12 February 2014 |
| Middle | Deir El Balah | 86,148.19 | 92,004.30 | 3:00 | 19 February 2014 |
| | Zawaida | 88,873.47 | 95,118.91 | 1:30 | 19 February 2014 |
| North | Furosia | 97,661.13 | 106,020.06 | 47:30 | 2–4 March 2014 |
| | Waha (two tests) | 99,830.23 | 108,689.72 | 9:30 | 1 March 2014 |
| | | | | 11:30 | 2 March 2014 |

*3.4. SGD Model from Radon Tracing*

Estimating groundwater discharge via radon is based on a mass balance approach. Inventories are measured, either as a snapshot or continuously over time, and these inventories are converted to input fluxes after making allowances for losses, due to decay, atmospheric evasion, and net coastal "mixing" terms. Although changing radon concentrations in coastal waters could be in response to a number of processes, the advective transport of radon-rich groundwater (pore water) through sediment is often the dominant process [37]. Thus, if one can measure or estimate the radon concentration in these advecting fluids, the $^{222}$Rn fluxes may be easily converted to water fluxes. $^{222}$Rn is also produced by $^{226}$Rn in the sampled water, and this quantity has to be subtracted to obtain "Excess $^{222}$Rn", which is relevant for our purposes. The inventory refers to the total amount of excess $^{222}$Rn ($^{222}$Rn in water minus $^{226}$Rn in pore water) per unit area. Decay is not considered because the fluxes are evaluated on a very short (1–2 h) time scales that are relative to the half-life of $^{222}$Rn [38].

The following data are required to perform the analysis: continuous measurement of total $^{222}$Rn activities (dpm·m$^{-3}$) in the coastal water column, water depth measurement, water and air temperatures, wind speed, atmospheric $^{222}$Rn concentrations, and $^{226}$Ra in the coastal water.

RAD7 device measures the total $^{222}$Rn (dpm·m$^{-2}$) in the water column every 30 min. $^{226}$Ra could not be measured during our research; hence, it has been taken from Weinstein et al. [39]. They measured $^{226}$Ra several times (Carmel coast in Israel, 60 km to the north of Gaza Strip) by gamma spectrometry, to show $^{226}$Ra activity in the range of 200–240 dpm/kg in the sand. On the other hand, they found $^{222}$Rn in pore water to be in a range of 340 to 390 dpm·L$^{-1}$ for the Kurkar groundwater [39].

Due to un-accessibility to measure the $^{222}$Rn activity offshore, we used the value of 1000 dpm·m$^{-3}$ taken from Peterson et al. [40]. The wind speed at the time of measurement was taken from http://www.wunderground.com.

We used the various assumptions that are inherent in the application of a $^{222}$Rn box model as stated in Burnett and Dulaiova [3], Dulaiova et al. [30] and Burnett et al. [38], to derive a rate of submarine groundwater discharge.

## 4. Results

### 4.1. Nutrients

High contents of (NO$_3^-$ + NO$_2^-$) in the coastal samples may be representing SGD (Figure 4A). The average groundwater concentration is 440 µM in wells, while in the coastal samples, the concentration ranges from 204.6 to 320.1 µM. A trend of decreasing (NO$_3^-$ + NO$_2^-$) (Table 2) is shown from the north to the south of Gaza Strip. Literature states that fresh groundwater concentration for (NO$_3^-$ + NO$_2^-$) is around 330 µM [11]. The analyzed concentrations in the coastal samples of our study, thus indeed indicating a substantial contribution of SGD.

The average concentration in the sampled groundwater wells was 4.9 µM of PO$_4^{3-}$, while the coastal samples were in the range of 5.6 to 11 µM (Table 2 and Figure 4B). Phosphate in the groundwater wells is lower than the measured phosphate in the coastal samples.

**Table 2.** The number of coastal shallow groundwater nutrient samples and average concentrations in different parts of the Gaza Strip.

| Area (Group) | Number of Samples | Statistical Parameter | NO$_3^-$ (µM) | NO$_2^-$ (µM) | NO$_3^-$ + NO$_2^-$ (µM) | NH$_4^+$ (µM) | PO$_4^{3-}$ (µM) | Salinity (ppt) |
|---|---|---|---|---|---|---|---|---|
| North | 51 | Average | 316.2 | 3.9 | 320.1 | 380.8 | 11.0 | 39.6 |
| | | Standard deviation | 167.4 | 4.3 | 167.1 | 152.4 | 7.0 | 0.5 |
| Middle | 68 | Average | 254.7 | 9.9 | 264.6 | 373.3 | 6.5 | 38.8 |
| | | Standard deviation | 117.3 | 17.7 | 130.4 | 223.3 | 5.2 | 0.7 |
| South | 109 | Average | 201.4 | 3.2 | 204.6 | 362.5 | 5.6 | 38.4 |
| | | Standard deviation | 80.9 | 6.8 | 83.5 | 198.5 | 5.1 | 1.7 |
| Wells (600 m from the coastline) | 4 | Average | 424.7 | 15.6 | 440.3 | 360 | 4.9 | 4.9 |
| | | Standard deviation | 262.0 | 30.0 | 276.0 | 201.8 | 6.6 | 4.9 |

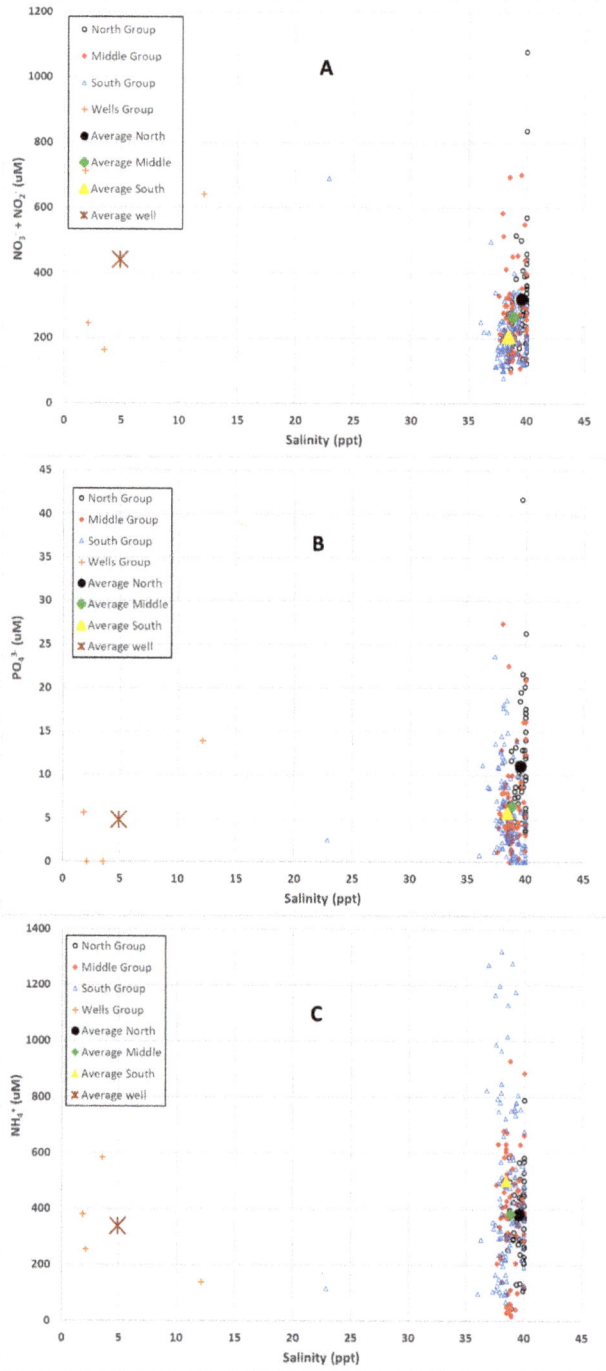

**Figure 4.** (**A**) Average $(NO_3^- + NO_2^-)$, (**B**) Average $PO_4^{3-}$ and (**C**) Average $NH_4^+$ vs salinity in Gaza coastal shallow groundwater and in deeper wells.

The average concentration of $NH_4^+$ in the onshore wells is 360 µM, while the coastal samples are ranging from 362.5 to 380.8 µM (Table 2 and Figure 4C). Hence, the fluxes to the sea have higher $NH_4^+$ concentrations than seawater that is poor in $NH_4^+$ (the ammonia concentrations in seawater vary from <0.1 to 15.3 µM [41]).

The nutrient analytical results for the three groups (north, middle, south) were tested against the significance of their difference, and it was found that, for nitrate, the three groups have significantly different averages at the 5% significance level. This was not always the case for the other parameters.

*4.2. Radon*

Seven measurements for continuous radon monitoring using RAD7 and RAD-AQUA were carried out in six locations (Figure 3 and Table 1).

The extracted data were processed following the Burnett and Dulaiova [3] model. The results processed by the model are summarized in Table 3, while Figure 5 shows the model results for the SGD flux per unit area based on each sample's time series. Figure 6 illustrates the relationship between the ebb/flood tide (water depth) and the SGD rate at the Furosia location, where positive advection fluxes represent groundwater discharge into the Mediterranean Sea and tend to occur during ebb tide.

In the north, three measurements have been carried out, one in Furosia and two in Waha sites. The Waha site should have one continuous measurement but due to pump failure after 9.5 h, we stopped the measurement and continued with a second measurement after replacing the portable submersible pump. The results show $0.92 \pm 0.27$ dpm·$L^{-1}$ for radon in water at the Furosia site, which results in 1.4 cm·$day^{-1}$ as an average SGD flux per unit area for two days of measurements. At the Waha site, radon was in the range of $0.37 \pm 0.17$ to $0.45 \pm 0.18$ dpm·$L^{-1}$ and the produced SGD rate is 1.5 cm·$day^{-1}$ as an average for both measurements. The average SGD rate in the north of Gaza Strip is 1.5 cm·$day^{-1}$.

In the south of the Gaza Strip, higher radon concentrations were found: $3.51 \pm 0.67$ and $8.2 \pm 1.2$ dpm·$L^{-1}$ in Rafah and Khan Younis, respectively, which produced SGD fluxes per unit area of 5.1 cm·$day^{-1}$ and 5.9 cm·$day^{-1}$ in Rafah and Khan Younis, respectively (average of 5.5 cm·$day^{-1}$).

The calculated SGD in the middle part of Gaza Strip showed identical values of 0.9 cm·$day^{-1}$ at both measuring sites, which gave us some confidence that the produced fluxes, in spite of the short time during which they were measured, were still fine. Further monitoring is recommended to enhance the results.

A sensitivity analysis was carried out to check the effect of the value for $^{222}$Rn offshore taken from Peterson et al. [40] (1000 dpm·$m^{-3}$) to the overall SGD rate calculations, by dividing and multiplying the value for $^{222}$Rn offshore by 2, and re-calculating the SGD rate for each measured location (Table 4).

Table 3. Results summary of continuous $^{222}$Rn measurements.

| Sample No. | Area (Group) | LOCATION NAME | No. of Records | Average Measured $^{222}$Rn (dpm·L$^{-1}$) | Average Depth (m) | Average Water Temperature (°C) | Calculated SGD Rate (cm·day$^{-1}$) | | |
|---|---|---|---|---|---|---|---|---|---|
| | | | | | | | Average | | Standard Deviation |
| 1 | North | Waha-1 | 20 | 0.45 ± 0.18 | 0.85 | 20.3 | 2.0 | | 1.8 |
| | | Waha-2 | 24 | 0.37 ± 0.17 | 0.83 | 20.6 | 1.0 | 1.5 | 0.9 |
| 2 | | Furosia | 96 | 0.92 ± 0.27 | 0.87 | 21.6 | 1.4 | | 1.3 |
| 3 | Middle | Zawaida | 3 | 0.17 ± 0.08 | 1.05 | 20.0 | 0.9 | | 1.3 |
| 4 | | Deir El Balah | 5 | 0.24 ± 0.11 | 0.88 | 21.2 | 0.9 | 0.9 | 0.8 |
| 5 | South | Khan Younis | 26 | 8.20 ± 1.20 | 0.89 | 18.5 | 5.9 | | 4.6 |
| 6 | | Rafah | 15 | 3.51 ± 0.67 | 0.93 | 19.0 | 5.1 | 5.5 | 3.7 |

**Table 4.** Sensitivity analysis for the effect of $^{222}$Rn offshore to the calculated submarine groundwater discharge (SGD) rate.

| Location Name | Final SGD Rate (cm·day$^{-1}$) with $^{222}$Rn Offshore (dpm·m$^{-3}$) | | |
|---|---|---|---|
| | 500 | 1000 | 2000 |
| Waha-1 | 2.2 | 2.0 | 1.9 |
| Waha-2 | 1.0 | 1.0 | 1.0 |
| Furosia | 1.4 | 1.4 | 1.3 |
| Zawaida | 0.8 | 0.9 | 1.0 |
| Deir El Balah | 0.8 | 0.9 | 1.1 |
| Khan Younis | 6.7 | 5.9 | 6.0 |
| Rafah | 5.1 | 5.1 | 5.1 |

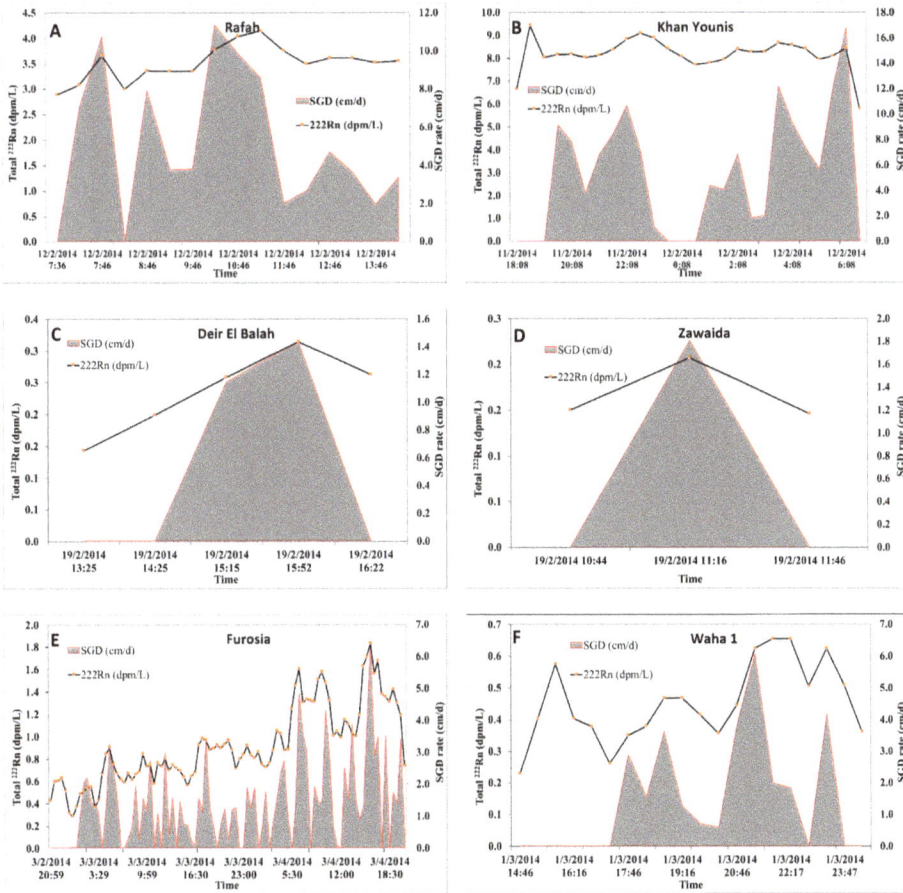

**Figure 5.** *Cont.*

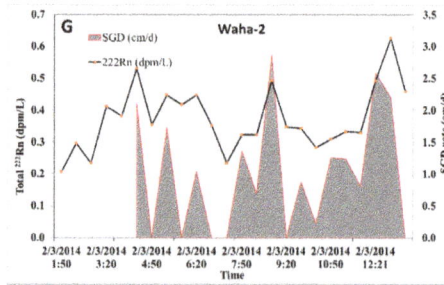

**Figure 5.** Radon in water and SGD rate in (**A**) Rafah, (**B**) Khan Younis, (**C**) Deir El Balah, (**D**) Zawaida, (**E**) Furosia, (**F**) Waha-1 and (**G**) Waha-2.

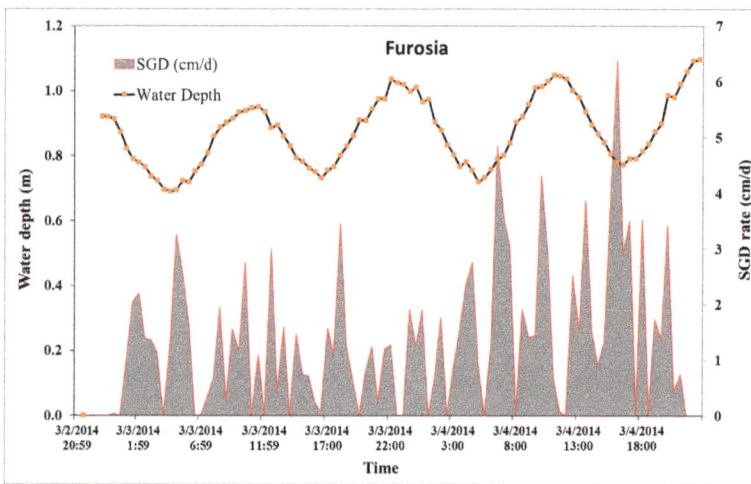

**Figure 6.** Relationship between ebb/flood tide and the SGD rates.

## 5. Discussion

This is the first study that has been carried out to quantify submarine groundwater discharge in the Gaza Strip, whereas seawater intrusion has been studied in this area, due to extensive abstraction and low groundwater recharge. Gaza Strip groundwater abstraction amounts to 170 million m$^3$ [24,42], while the long term rainfall recharge is 24.4 million m$^3$/year [21], leading to lowering of the groundwater level by more than 10 m in the south, and 5 m in the north.

Nutrient analysis is a good way to identify potential locations of SGD occurrence, taking into consideration any other nutrient source (i.e., sewage outflow to the sea and/or agricultural activities close to the sea). Seawater is mainly poor in nutrients; hence, any high level of nutrients along the shoreline would be from groundwater. In this study, we used the nutrient analysis to indicate the potential sites for SGD, and to identify potential radon measuring sites. We found that the average ($NO_3^- + NO_2^-$) in all groups had increased values. Also, considering the high $NH_4^+$ concentration in coastal samples compared to the seawater to be within the range of the onshore groundwater wells, this allowed us to conclude that SGD was indeed taking place.

Phosphate is as low as 5 µM in the groundwater wells, probably due to phosphate removal by water–rock interaction, while the coastal samples showed higher phosphate concentration in all groups, except the samples in Rafah governorate.

Only one method of SGD quantification has been applied in the study area, while other tools could not be carried out to enhance the accuracy of the outcomes. Seepage meters collecting in situ pore water in a closed system, are expected to provide more reliable results on the local scale.

The results show a variation in SGD quantity from the north to the south of Gaza Strip, where the southern part has higher fluxes than the north, while the middle area of Gaza Strip has a low quantity of SGD.

In the northern part of the Gaza Strip, where three continuous radon measurements were performed, the SGD rates were in the range of 1.0 to 2.0 cm·day$^{-1}$ (average of 1.5 cm·day$^{-1}$). On the other hand, the SGD rates were higher in the south (range of 5.1 to 5.9 cm·day$^{-1}$), while they were minimal in the middle part (0.9 cm·day$^{-1}$). It was calculated (Table 5) that, considering solely a strip of 1 m wide along the coastline, in the north, a groundwater quantity of $97 \times 10^3$ m$^3$/year is discharged into the Mediterranean Sea, while this is only $31 \times 10^3$ m$^3$/year in the middle part, and a large flux of $247 \times 10^3$ m$^3$/year in the south. The results thus show a variation in SGD quantity from the north to the south of the Gaza Strip, where the southern part shows higher fluxes than the north, while the middle area shows low quantities of SGD.

**Table 5.** SGD and associated nutrient discharge occurring in a 1 m wide strip along the Gaza coast.

| Area (Group) | Coastline Length (km) | SGD Rate cm/day | SGD Quantity m$^3$/year | NO$_3^-$ | NO$_2^-$ | NH$_4^+$ | PO$_4^{3-}$ |
|---|---|---|---|---|---|---|---|
| North | 18.1 | 1.5 | 97,116 | 316.2 µM | 3.9 µM | 380.8 µM | 11 µM |
| | | | | 1412.6 kg/year | 23.5 kg/year | 665.7 kg/year | 101.5 kg/year |
| Middle | 9.5 | 0.9 | 31,208 | 254.7 µM | 9.9 µM | 373.3 µM | 6.5 µM |
| | | | | 365.6 kg/year | 19.2 kg/year | 209.7 kg/year | 19.3 kg/year |
| South | 12.3 | 5.5 | 246,923 | 201.4 µM | 3.2 µM | 362.5 µM | 5.6 µM |
| | | | | 2287.6 kg/year | 49.0 kg/year | 1611.2 kg/year | 131.4 kg/year |
| Total for all Gaza Strip coastline (1 m wide strip) | | | 375,246 | 4066 kg/year | 92 kg/year | 2487 kg/year | 252 kg/year |

Calculating SGD quantity in a 1 m wide strip along the whole Gaza coastline results in total SGD of 0.38 million m$^3$/year, where the SGD quantity in the south is more than two times higher compared to the north. SGD may be expected to occur over a much wider area, although the rates can be expected to decrease with increasing distance from the coast. For this reason, we refrain from taking the whole outcrop area of Kurkar Group at the sea bottom into account, as the measured SGD may be expected to represent the maximum rates, which are not representative for the whole submarine outcrop area. Considering that SGD would occur with this rate in a strip of 100 m wide (which still appears to be a conservative estimate), results in a quantity of 38 million m$^3$ of groundwater being discharged yearly to the Mediterranean Sea along Gaza coast.

The reason behind having higher SGD in the south compared to the north, while the recharge rate from rainfall is higher in the north than in the south [21], probably refers to the amount of abstraction quantities (in the north, 61.6% of total groundwater abstraction (for domestic water supply) is located, while it is only 16.4% in the middle area and 22% in the southern part of Gaza Strip; [24]). More abstraction in the north leads to less groundwater discharge to the sea. On the other hand, the low SGD rate in the middle part of Gaza Strip (0.9 cm·day$^{-1}$ on average) is probably related to the presence of Sabkhas, which act as discharge areas for groundwater and intercept the groundwater before it can reach the sea.

In a coastal strip of only 1 m wide, a yearly discharge of over 4 tons of nitrate and 2.5 tons of ammonium to the Mediterranean Sea occurs (Table 5). Assuming that SGD with the measured rates would occur in a strip of 100 m wide, would result in nutrient discharges that are one hundred times larger than calculated in Table 5, that is, over 400 tons of nitrate and 250 tons of ammonium that are yearly discharged by Gaza Strip groundwater into the sea.

The obtained results from using different $^{222}$Rn offshore values (Table 4) showed no significant difference to the overall calculated SGD rate, which gave us confidence to use $^{222}$Rn offshore, as stated in Peterson et al. [40].

## 6. Conclusions

Using nutrient analytical results along the shoreline helps us to identify potential sites of submarine groundwater discharge, after eliminating sites where nutrients may be due to other potential sources (such as sewage outflow to the shoreline and/or agricultural practices that are close to the shoreline). In our study, the raised nutrient concentrations in shallow groundwater collected near the shoreline, helped us to confirm that SGD is taking place, and that it can be further investigated by the means of $^{222}$Rn measurements.

Gaza Strip has suffered from seawater intrusion in the past few decades, even though groundwater discharge to the Mediterranean Sea is still occurring, with a range of up to 5.9 cm·day$^{-1}$ (in the selected sites the minimum SGD rate was 0.9 cm·day$^{-1}$). SGD rate was found to be maximal in the south of Gaza Strip (average of 5.5 cm·day$^{-1}$), while being much lower in the north (average of 1.5 cm·day$^{-1}$) in spite of the substantially higher groundwater recharge in the north. This must be related to groundwater exploitation, which is much higher in the north. In central Gaza, measured SGD rate is minimal (average of 0.9 cm·day$^{-1}$), which must be due to the presence of Sabkhas, acting as groundwater discharge areas, and intercepting the groundwater before it reaches the sea. The sensitivity analysis carried out for $^{222}$Rn offshore showed no major effect on the calculated SGD rate in our study.

The analysis showed a groundwater discharge to the sea of about 0.38 million m$^3$/year in a 1 m wide strip along the full length of Gaza coast. Assuming that the same SGD rate occurs in a 100 m wide strip, this results in a yearly SGD quantity of 38 million m$^3$/year along the Gaza coastline. This SGD transports over 400 tons of nitrate and 250 tons of ammonium to the Mediterranean Sea.

Regular measurements of SGD will enhance our understanding of water balance in the Gaza Strip, and describe the temporal and spatial variability of SGD along the Gaza coastal aquifer.

**Author Contributions:** Writing—review & editing, A.M. and K.W.

**Funding:** This research received no external funding.

**Acknowledgments:** The authors thank Jehad Dasht, and Ramadan Murtaja for their assistance in the field, while continuous radon measurements were performed. We thank Fadi Abo Shanab for his assistance in analyzing the nutrient samples. We are grateful to the Global Minds Fund of Ghent University for providing some funding for a research stay for Ashraf Mushtaha in Ghent. The authors want to thank the two anonymous reviewers, whose constructive criticism contributed to substantially improving the paper.

**Conflicts of Interest:** The authors declare no conflicts of interest.

## References

1. Zektzer, I.S.; Ivanov, V.A.; Meskheteli, A.V. The problem of direct groundwater discharge to the seas. *J. Hydrol.* **1973**, *20*, 1–36. [CrossRef]
2. Moore, W.S. The subterranean estuary: A reaction zone of ground water and sea water. *Mar. Chem.* **1999**, *65*, 111–125. [CrossRef]
3. Burnett, W.C.; Dulaiova, H. Estimating the dynamics of groundwater input into the coastal zone via continuous radon-222 measurements. *J. Environ. Radioact.* **2003**, *69*, 21–35. [CrossRef]
4. Buddemeier, R.W. Groundwater flux to the ocean: Definitions, data, applications, uncertainties. In *Groundwater Discharge in the Coastal Zone: Proceedings of an International Symposium*; Buddemeier, R.W., Ed.; IGBP: Texel, The Netherlands, 1996; pp. 16–21.
5. Dollar, S.J.; Atkinson, M.J. Effects of nutrient subsidies from groundwater to nearshore marine ecosystems off the island of Hawaii. *Estuar. Coast. Shelf Sci.* **1992**, *35*, 409–424. [CrossRef]
6. Paerl, H.W. Coastal eutrophication and harmful algal blooms: Importance of atmospheric deposition and groundwater as "new" nitrogen and other nutrient sources. *Limnol. Oceanogr.* **1997**, *42*, 1154–1165. [CrossRef]

7.  Miller, D.C.; Ullman, W.J. Ecological consequences of ground water discharge to Delaware Bay, United States. *Groundwater* **2004**, *42*, 959–970. [CrossRef]
8.  Dailer, M.L.; Ramey, H.L.; Saephan, S.; Smith, C.M. Algal δ15N values detect a wastewater effluent plume in nearshore and offshore surface waters and three dimensionally model the plume across a coral reef on Maui, Hawai'i, USA. *Mar. Pollut. Bull.* **2012**, *64*, 207–213. [CrossRef]
9.  Amato, D.W.; Bishop, J.M.; Glenn, C.R.; Dulai, H.; Smith, C.M. Impact of submarine groundwater discharge on marine water quality and reef biota of Maui. *PLoS ONE* **2016**, *11*, e0165825. [CrossRef]
10. Kelly, J.L.; Dulai, H.; Glenn, C.R.; Lucey, P.G. Integration of aerial infrared thermography and in situ radon-222 to investigate submarine groundwater discharge to Pearl Harbor, Hawaii, USA. *Limnol. Oceanogr.* **2018**, *9999*, 1–20. [CrossRef]
11. Weinstein, Y.; Yechieli, Y.; Shalem, Y.; Burnett, W.C.; Swarzenski, P.W.; Herut, B. What Is the Role of Fresh Groundwater and Recirculated Seawater in Conveying Nutrients to the Coastal Ocean? *Environ. Sci. Technol.* **2011**. [CrossRef]
12. Lebbe, L.C. The subterranean flow of fresh and salt water underneath the western Belgian beach. Proceedings of the Seventh Salt Water Intrusion Meeting, Uppsala. *Sver. Geol. Unders. Rap. Medd.* **1981**, *27*, 193–219.
13. Werner, A.D.; Lockington, D.A. Tidal impacts on riparian salinities near estuaries. *J. Hydrol.* **2006**, *328*, 511–522. [CrossRef]
14. Bishop, J.M.; Glenn, C.R.; Amato, D.W.; Dulai, H. Effect of land use and groundwater flow path on submarine groundwater discharge nutrient flux. *J. Hydrol. Reg. Stud.* **2017**, *11*, 194–218. [CrossRef]
15. Burnett, W.C.; Cable, J.E.; Corbett, D.R.; Chanton, J.P. Tracing groundwater flow into surface waters using natural 222Rn. In *Groundwater Discharge in the Coastal Zone: Proceedings of an International Symposium*; Buddemeier, R.W., Ed.; IGBP: Texel, The Netherlands, 1996; pp. 22–28.
16. Cable, J.E.; Burnett, W.C.; Chanton, J.P.; Weatherly, G. Modeling groundwater flow into the ocean based on 222Rn. *Earth Planet. Sci. Lett.* **1996**, *144*, 591–604. [CrossRef]
17. Corbett, D.R.; Burnett, W.C.; Cable, P.H. Tracing of groundwater input into Par Pond, Savannah River Site by Rn-222. *J. Hydrol.* **1997**, *203*, 209–227. [CrossRef]
18. Burnett, W.; Chariton, J.; Christoff, J.; Kontar, E.; Krupa, S.; Lambert, M.; Moore, W.; O'Rourke, D.; Paulsen, R.; Smith, C.; et al. Assessing methodologies for measuring groundwater discharge to the ocean. *EOS* **2002**, *83*, 117, 122–123. [CrossRef]
19. Lane-Smith, D.R.; Burnett, W.C.; Dulaiova, H. Continuous radon-222 measurements in the coastal zone. *Sea Technol.* **2002**, *43*, 37–45.
20. Burnett, W.C.; Kim, G.; Lane-Smith, D. A continuous radon monitor for assessment of radon in coastal ocean waters. *J. Radioanal. Nucl. Chem.* **2001**, *249*, 167–172.
21. Mushtaha, A.; Van Camp, M.; Walraevens, K. Evolution of runoff and groundwater recharge in the Gaza Strip over the four decades. *Environ. Earth Sci.* **2018**. under review.
22. PWA/USAID. *Integrated Aquifer Management Plan*; PWA: Gaza Strip, Palestine, 2000.
23. Mushtaha, A.; Aliewi, A.; Mackay, R. The Use of Scavenger Wells to Control Saltwater Upconing in Gaza, Palestine. In Proceedings of the 16th Salt Water Intrusion Meeting, Miedzyzdroje–Wolin Island, Poland, 12–15 June 2000; pp. 109–116.
24. PWA. *Water Resources Status Summary Report/Gaza Strip*; Water Resources Directorate: Palestinian Water Authority, Palestine, 2017.
25. PWA. *Water Resources Status Report for Year 2014/Gaza Strip*; Water Resources Directorate: Palestinian Water Authority, Palestine, 2015.
26. Greitzer, Y.; Dan, J. *The Effect of Soil Landscape and Quaternary Geology on the Distribution of Saline and Fresh Water Aquifers in the Coastal Plain of Israel*; Tahal Water Planning for Israel, Ltd.: Tel Aviv, Israel, 1967.
27. Kress, N.; Herut, B. Spatial and seasonal evolution of dissolved oxygen and nutrients in the Southern Levantine Basin (Eastern Mediterranean Sea): Chemical characterization of the water masses and inferences on the high N:P ratio. *Deep Sea Res.* **2001**, *48*, 2347–2372. [CrossRef]
28. Dulaiova, H.; Gonneea, M.E.; Henderson, P.B.; Charette, M.A. Geochemical and physical sources of radon variation in a subterranean estuary—Implications for groundwater radon activities in submarine groundwater discharge studies. *Mar. Chem.* **2008**, *110*, 120–127. [CrossRef]
29. Dulaiova, H.; Burnett, W.C. Radon loss across the water-air interface (Gulf of Thailand) estimated experimentally from $^{222}$Rn-$^{224}$Ra. *Geophys. Res. Lett.* **2006**, *33*. [CrossRef]

30. Dulaiova, H.; Burnett, W.C.; Chanton, J.P.; Moore, W.S.; Bokuniewicz, H.J.; Charette, M.A.; Sholkovitz, E. Assessment of Submarine Groundwater Discharges into West Neck Bay, New York, via Natural Tracers. *Cont. Shelf Res.* **2006**, *26*, 1971–1983. [CrossRef]

31. Christoff, J.L. Quantifying Groundwater Seepage into a Shallow Near-Shore Coastal Zone by Two Techniques. Master's. Thesis, Florida State University, Tallahassee, FL, USA, 2001.

32. Dulaiova, H.; Burnett, W.C.; Wattayakorn, G.; Sojisuporn, P. Are Groundwater Inputs into River-Dominated Areas Important? The Chao Phraya River—Gulf of Thailand. *Limnol. Oceanogr.* **2006**, *51*, 2232–2247. [CrossRef]

33. Mulligan, A.E.; Charette, M.A. Intercomparison of submarine groundwater discharge estimates from a sandy unconfined aquifer. *J. Hydrol.* **2006**, *327*, 411–425. [CrossRef]

34. Chidambaram, S.; Nepoliana, M.; Ramanathan, A.L.; Sarathidasan, J.; Thilagavathi, R.; Thivya, C.; Prasanna, M.V.; Srinivasamoorthy, K.; Jacob, N.; Mohokar, H. An attempt to identify and estimate the subsurface groundwater discharge in the south east coast of India. *Int. J. Sustain. Built Environ.* **2017**, *6*, 421–433. [CrossRef]

35. Broecker, W.S. An application of natural radon to problems in oceanic circulation. In *Symposium on Diffusion in the Oceans and Fresh Waters*; Ichiye, D.T., Ed.; Lamont Geological Observatory: New York, NY, USA, 1965; pp. 116–145.

36. Mathieu, G.; Biscayne, P.; Lupton, R.; Hammond, D. System for measurements of 222Rn at low levels in natural waters. *Health Phys.* **1988**, *55*, 989–992. [CrossRef]

37. Ray, R.L.; Dogan, A. Contemporary Methods for Quantifying Submarine Groundwater Discharge to Coastal Areas. In *Emerging Issues in Groundwater Resources. Advances in Water Security*; Fares, A., Ed.; Springer: Cham, Switzerland, 2016. [CrossRef]

38. Burnett, W.C.; Santos, I.R.; Weinstein, Y.; Swarzenski, P.W.; Herut, B. Remaining uncertainties in the use of Rn-222 as a quantitative tracer of submarine groundwater discharge. In Proceedings of the International Symposium: A New Focus on Groundwater-Seawater Interactions, Perugia, Italy, 2–13 July 2007; pp. 109–118.

39. Weinstein, Y.; Burnett, W.C.; Swarzenski, P.W.; Shalem, Y.; Yechieli, Y.; Herut, B. Role of aquifer heterogeneity in fresh groundwater discharge and seawater recycling: An example from the Carmel coast, Israel. *J. Geophys. Res.* **2007**, *112*, C12016. [CrossRef]

40. Peterson, N.R.; Burnett, W.C.; Taniguchi, M.; Chen, J.; Santos, I.R.; Ishitobi, T. Radon and radium isotope assessment of submarine groundwater discharge in the Yellow River delta, China. *J. Geophys. Res.* **2008**, *113*, C09021. [CrossRef]

41. Hampson, B.L. The Analysis of Ammonia in Polluted sea water. *Water Res.* **1977**, *11*, 305–308. [CrossRef]

42. CMWU. *Water Resource Status in the Gaza Strip*; Coastal Municipalities Water Utility: Al-Zahraa City, Palestine, 2013.

*water*

MDPI

*Article*

# Comparative Evaluation of ANN- and SVM-Time Series Models for Predicting Freshwater-Saltwater Interface Fluctuations

Heesung Yoon [1], Yongcheol Kim [1,*], Kyoochul Ha [1], Soo-Hyoung Lee [1] and Gee-Pyo Kim [2]

[1] Korea Institute of Geoscience and Mineral Resources, Daejeon 34132, Korea; hyoon@kigam.re.kr (H.Y.); hasife@kigam.re.kr (K.H.); rbagio@kigam.re.kr (S.-H.L.)

[2] Jeju Water Resources Headquarters, Jeju 63343, Korea; andwater@korea.kr

* Correspondence: yckim@kigam.re.kr; Tel.: +82-42-868-3086

Academic Editors: Maurizio Polemio and Kristine Walraevens
Received: 21 February 2017; Accepted: 2 May 2017; Published: 4 May 2017

**Abstract:** Time series models based on an artificial neural network (ANN) and support vector machine (SVM) were designed to predict the temporal variation of the upper and lower freshwater-saltwater interface level (FSL) at a groundwater observatory on Jeju Island, South Korea. Input variables included past measurement data of tide level (T), rainfall (R), groundwater level (G) and interface level (F). The T-R-G-F type ANN and SVM models were selected as the best performance model for the direct prediction of the upper and lower FSL, respectively. The recursive prediction ability of the T-R-G type SVM model was best for both upper and lower FSL. The average values of the performance criteria and the analysis of error ratio of recursive prediction to direct prediction (RP-DP ratio) show that the SVM-based time series model of the FSL prediction is more accurate and stable than the ANN at the study site.

**Keywords:** artificial neural network; support vector machine; time series model; freshwater-saltwater interface; direct prediction; recursive prediction

---

## 1. Introduction

Monitoring and forecasting of temporal changes of the freshwater-saltwater interface level (FSL) in coastal areas is necessary for the early detection of saltwater intrusion and the management of coastal aquifers. To measure the location and variation of FSL, the geophysical well logging technique to capture the vertical profile of electrical conductivity or salinity has been traditionally employed [1,2]. Recently, research has been conducted on the development of an interface egg device to monitor the temporal variation of FSL [3].

For the simulation or prediction of saltwater intrusion into aquifers, physics-based numerical models have been developed and applied to various field sites; Gingerich and Voss [4] applied 3D-SUTRA model to a coastal aquifer in Hawaii for the simulation of the saltwater intrusion; Werner and Gallagher [5] characterized seawater intrusion in coastal aquifers of the Pioneer Valley, Australia using MODHMS model; Guo and Langevin [6] developed SEAWAT, a variable-density finite-difference groundwater flow mode and Rozell and Wong [7] applied it to Shelter Island, USA for assessing effects of climate change on the groundwater resources; Yechieli et al. [8] examined the response of the Mediterranean and Dead Sea coastal aquifers using FEFLOW model. Physics-based numerical models are powerful tools for the simulation or prediction of temporal and spatial variation of FSL in a given domain. However, they require a large quantity of precise data related to the physical properties of the domain, a lack of which can cause severe deterioration in the accuracy and reliability of their results [9,10]. Time series modeling can be an effective alternative approach for predicting saltwater intrusion where geological and geophysical surveys are limited and monitoring data of temporal variation related to saltwater intrusion are available.

Recently, in the field of hydrology and hydrogeology, research on the application of time series models—based on machine learning techniques such as an artificial neural network (ANN) and a support vector machine (SVM) to prediction of water resources variations—have been increased; Zealand et al. [11] utilized the ANN for forecasting short term stream flow of the Winnipeg River system in Canada; Akhtar et al. [12] applied ANN to river flow forecasting at Ganges river; Hu et al. [13] explored new measures for improving the generalization ability of the ANN for the prediction of the rainfall-runoff; Coulibaly et al. [14] and Mohanty et al. [15] examined the performance of ANN for the prediction of groundwater level (GWL) fluctuations; Coppola et al. [16] used the ANN for the prediction of GWL under variable pumping conditions; Liong and Sivapragasam [17], and Yu et al. [18] employed the SVM for the prediction of the flood stage; Asefa et al. [19] used the SVM for designing GWL monitoring networks; Gill et al. [20] assessed the effect of missing data on the performance of ANN and SVM models for GWL prediction; Yoon et al. [21] used ANN and SVM for long-term GWL forecast.

For coastal aquifer management, time series models have been developed to predict groundwater level (GWL) fluctuations using machine learning methods [22–24]. In the domain of saltwater intrusion, recent studies have used a time series modeling approach [25,26]; however, their target was to predict salinity at coastal rivers rather than FSL change in coastal aquifers.

In this study, we monitored temporal variations of the upper and lower FSL of a groundwater observatory at Jeju Island in South Korea. Using the observed FSL data, we designed time series models based on artificial neural networks and support vector machines for the prediction of FSL fluctuations. The prediction accuracy of FSL was estimated with different structures of models. The paper is organized as follows: Section 2 describes the study site and FSL monitoring data. Section 3 describes the development of the time series models for FSL prediction based on artificial neural networks and the support vector machines. The FSL prediction results are described and discussed in Section 4, and conclusions are drawn in Section 5.

## 2. FSL Monitoring

### 2.1. Study Area

The study site is a groundwater observatory (HD2) located at the north-eastern part of Jeju Island in South Korea (Figure 1). Jeju Island is the largest volcanic island in South Korea with an area of 1849 km$^2$, where the mean annual air temperature is 16.2 °C and total annual precipitation is 1710 mm.

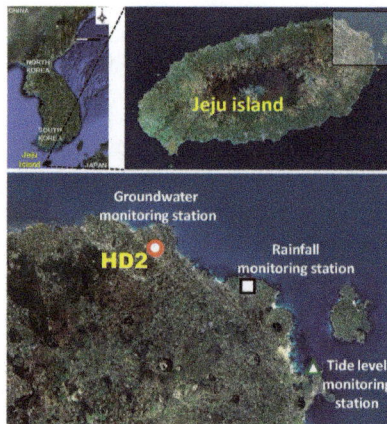

**Figure 1.** Location of the study site.

Perennial rivers are scarce on Jeju Island, so groundwater is a main water source for domestic, agricultural or industrial uses, as well as a main source of drinking water, therefore, the groundwater has been systematically managed with a saltwater intrusion monitoring network by the local government. The HD2 groundwater observatory (one of the saltwater intrusion monitoring networks), is located 2.3 km from the coast-line and 42.73 m (above mean sea level: AMSL). Rainfall and tide level monitoring stations are located near the north-eastern shoreline at a distance of 6.1 km and 12.3 km from HD2, respectively. A geophysical survey was conducted to measure the vertical profile of electrical conductivity in HD2. The results show that the freshwater-saltwater interface appears between −49.0 m and −62.0 m (AMSL) where the electrical conductivities are 16.2 mS/cm and 40.7 mS/cm, respectively (Figure 2).

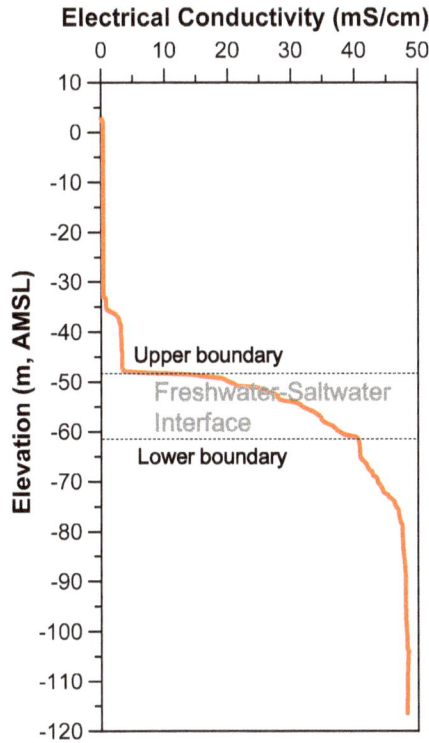

**Figure 2.** Vertical profile of electrical conductivity at the HD2 observatory.

*2.2. Monitoring Device and Data*

We utilized the interface egg developed by Kim et al. [3] to monitor the temporal variation of FSL at the HD2 observatory. The interface egg is a monitoring probe designed to have a specific density of the value between freshwater and saltwater, which enables it to float on the FSL based on the concept of neutral buoyancy. Using the measured pressure data of the interface egg, and a pressure sensor at a fixed depth, the position of the FSL at time $t$ is estimated as follows (Figure 3):

$$FSL(t) = EL - (b - a(t) + c(t)) \tag{1}$$

where EL is an elevation of a top of well casing; $a(t)$ is the pressure value measured at fixed depth $b$; and $c(t)$ is the pressure value measured at the interface egg.

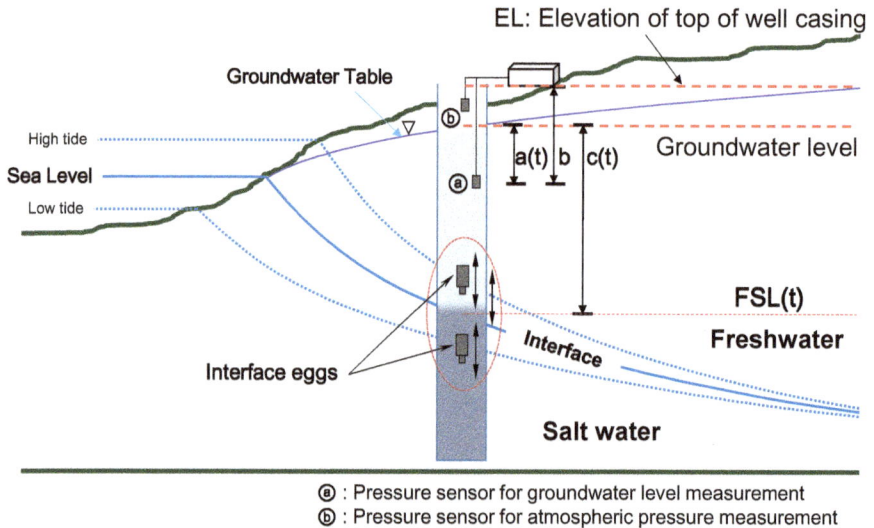

**Figure 3.** Schematic diagram of the freshwater-saltwater interface level (FSL) monitoring system using the interface egg (modified from Kim et al. [3]).

Taking into account the vertical profile of electrical conductivity, we installed two interface eggs at around −49.0 m and −62.0 m (AMSL) which corresponded to the upper and lower boundaries of the freshwater-saltwater interface at the HD2 observatory. We additionally installed a pressure sensor at a fixed depth to monitor the GWL fluctuations. Hourly measured data of GWL, upper and lower FSLs, rainfall and tide level from 15 September–5 October 2014, are shown in Figure 4.

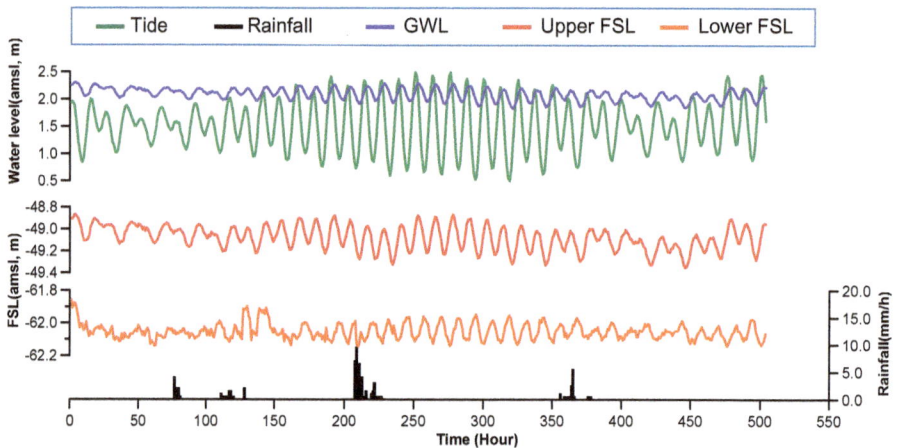

**Figure 4.** Time series data of groundwater level (GWL), upper and lower FSL, rainfall and tide level at the HD2 observatory.

The result of cross correlation analyses between the time series data at the HD2 observatory shows that the correlation of upper FSL with tide and GWL is much higher than of the lower FSL (Table 1). The maximum correlation coefficient between GWL and upper FSL is the highest: 0.97 at a lag time of

0 h, which indicates that the movement of the upper FSL is strongly and immediately influenced by GWL. Furthermore, the maximum correlation coefficient of tide–GWL and tide–upper FSL are as high as 0.85 and 0.83 at a lag time of 2 h, respectively. The correlation of rainfall with GWL and FSL is not significant for this study.

**Table 1.** Results of cross correlation analysis for measured time series data at the HD2 observatory.

| Variables | Max. Correlation Coefficient | Lag Time (Hour) |
|:---:|:---:|:---:|
| T-G | 0.85 | 2 |
| R-G | 0.14 | 43 |
| T-F (upper) | 0.83 | 2 |
| R-F (upper) | 0.11 | 47 |
| G-F (upper) | 0.97 | 0 |
| T-F (Lower) | 0.56 | 6 |
| R-F (Lower) | 0.27 | 19 |
| G-F (Lower) | 0.54 | 4 |

Notes: Where **T**: tide; **R**: rainfall; **G**: groundwater level; **F**: interface level.

## 3. FSL Prediction Model Development

We employed artificial neural network (ANN) and support vector machine (SVM) techniques to construct time series models for the prediction of the upper and lower FSL fluctuations. Theoretical backgrounds of the ANN, SVM, and time series modeling process are described below.

### 3.1. Aritificial Neural Network (ANN)

The ANN is a mathematical framework patterned after the parallel processing sequence of the human brain. A feedforward network (FFN), one of the most common structures of the ANN, is generally composed of three layers of input, hidden and output (Figure 5a). Each layer of the ANN has a certain number of nodes and each node in a layer is connected to other nodes in the next layer with a specific weight and bias. The mathematical expression of the calculation process in the FFN is as follows:

$$x_j = f \left( \sum_{i=1}^{n} w_{ij} x_i + b_j \right) \tag{2}$$

where the subscript $i$ and $j$ denote the previous and present layer, respectively; $x$ is the nodal value; $w$ and $b$ are weight and bias values, respectively; $n$ is the number of nodes in the previous layer; and $f$ denotes a transfer function of the present layer. Log-sigmoid and linear functions were allocated to hidden and output layers, respectively, which are known to be an effective combination for enhancing the extrapolation ability of the ANN [27,28].

The purpose of the ANN model building is to find the optimal values of weights and biases by learning or training from the given input and output data. We employed a back-propagation algorithm (BPA) with momentum suggested by Rumelhart and McClelland [29] for training the ANN. The weight and bias update rule of the BPA can be expressed as follows:

$$E^n = \sum_{k=1}^{N} (\hat{y}_k^n - y_k^n)^2 \tag{3}$$

$$\Delta w^n = \text{MM} \, \Delta w^{n-1} + (1 - \text{MM}) \, \text{LR} \left( -\frac{\partial E^n}{\partial w^n} \right) \tag{4}$$

$$\Delta b^n = \text{MM} \, \Delta b^{n-1} + (1 - \text{MM}) \, \text{LR} \left( -\frac{\partial E^n}{\partial b^n} \right) \tag{5}$$

where $E^n$ is a sum of squared errors between observed ($y$) and estimated ($\hat{y}$) values at $n$-th weight and bias update stage, MM and LR denote the momentum and learning rate values, respectively, and $N$ is the number of data allocated to the training stage. In this study, three model parameters; i.e., number of hidden nodes (HN), MM and LR were determined by a grid search that is one of the trial and error method. We took into account 6 values for every ANN model parameters: $HN \in [2, 5, 10, 15, 20, 25]$, $MM \in [0.0, 0.1, 0.3, 0.5, 0.7, 0.9]$, and $LR \in [0.001, 0.005, 0.01, 0.015, 0.02, 0.025]$, which composes 216 candidate groups of model parameters.

*3.2. Suport Vector Machine (SVM)*

The SVM, a relatively new machine learning method suggested by Vapnik [30], is based on the structural risk minimization (SRM) rather than the empirical risk minimization (ERM) of the ANN. From a data classification point of view, ERM based machine learning method is designed to minimize the error of the estimated classifier for the data in the training stage. Therefore, the model update is stopped when the error of the training stage data is zero or within a certain value of the tolerance. The SRM based method, such as the SVM, is designed to maximize a margin between data groups to be classified, which maximize the generalization ability of the model. The mathematical expression of the output estimation of the SVM is as follows:

$$S(\mathbf{x}) = \mathbf{w} \cdot \phi(\mathbf{x}) + b \tag{6}$$

where $S$ denotes an SVM estimator, $\mathbf{w}$ denotes a weight vector, $\phi$ is a nonlinear transfer function that maps input vectors into a higher-dimensional feature space. Platt [31] introduced a convex optimization problem with an $\varepsilon$-insensitivity loss function to find the solution of Equation (6) as follows:

$$
\begin{aligned}
&\underset{\mathbf{w},b,\zeta,\zeta*}{\text{minimize}} && \tfrac{1}{2}\|\mathbf{w}\|^2 + C \sum_{k=1}^{N} \left( \zeta_k + \zeta_k^* \right) \\
&\text{subject to} && \left\{
\begin{array}{l}
y_k - \mathbf{w}^T \phi(\mathbf{x}_k) - b \le \varepsilon + \zeta_k \\
\mathbf{w}^T \phi(\mathbf{x}_k) + b - y_k \le \varepsilon + \zeta_k^* \\
\zeta_k, \zeta_k^* \ge 0
\end{array}
\right\} \quad k = 1, 2, \cdots, N
\end{aligned}
\tag{7}
$$

where $\zeta$ and $\zeta^*$ are slack variables that penalize errors of estimated values over the error tolerance $\varepsilon$, $C$ is a trade-off parameter that controls the degree of the empirical error in the model building process, and $\mathbf{x}$ is the input vector in the training stage. Equation (7) can be solved using Lagrangian multipliers and the Karush-Kuhn-Tucker (KKT) optimality condition as follows:

$$
\begin{aligned}
&\underset{\alpha, \alpha^*}{\text{maximize}} && \left\{
\begin{array}{l}
-\tfrac{1}{2} \sum_{k,l=1}^{N} \left( \alpha_k - \alpha_k^* \right) \left( \alpha_l - \alpha_l^* \right) K(\mathbf{x}_k, \mathbf{x}_l) \\
-\varepsilon \sum_{k=1}^{N} \left( \alpha_k + \alpha_k^* \right) + \sum_{k=1}^{N} y_k \left( \alpha_k - \alpha_k^* \right)
\end{array}
\right. \\
&\text{subject to} && \left\{
\begin{array}{l}
\sum_{k=1}^{N} \left( \alpha_k - \alpha_k^* \right) = 0 \\
0 \le \alpha_k, \alpha_k^* \le C
\end{array}
\right. \\
&\text{to obtain} && F(\mathbf{x}) = \sum_{k=1}^{n} \left( \alpha_k - \alpha_k^* \right) K(\mathbf{x}, \mathbf{x}_k) + b
\end{aligned}
\tag{8}
$$

where $\alpha$ and $\alpha^*$ are Lagrangian multipliers, $K$ is a kernel function defined by an inner product of the nonlinear transfer functions. A radial basis function with parameter $\sigma$ is commonly used as the kernel function:

$$K(\mathbf{x}, \mathbf{x}_k) = \exp\left( -\frac{\|\mathbf{x} - \mathbf{x}_k\|^2}{2\sigma^2} \right) \tag{9}$$

We employed the sequential minimal optimization (SMO) algorithm [32] to solve Equation (8) and construct the SVM model. The SMO minimizes a subset of target variables by two and finds an analytical solution of the subset repeatedly, until all given input vectors satisfy the KKT conditions. A detailed explanation of the SMO algorithm can be found in References [31,32]. The SVM model parameters of $C$, $\varepsilon$, and $\sigma$ were selected by the grid search method like the ANN. We took into account six values for every SVM model parameters: $C \in [0.5, 1.0, 3.0, 5.0, 7.0, 10.0]$, $\varepsilon \in [0.01, 0.05, 0.1, 0.11, 0.12, 0.13]$, and $\sigma \in [0.5, 1.0, 1.5, 2.0, 2.5, 3.0]$, which composes 216 candidate combinations of model parameters. A schematic diagram of the SVM structure is shown in Figure 5b.

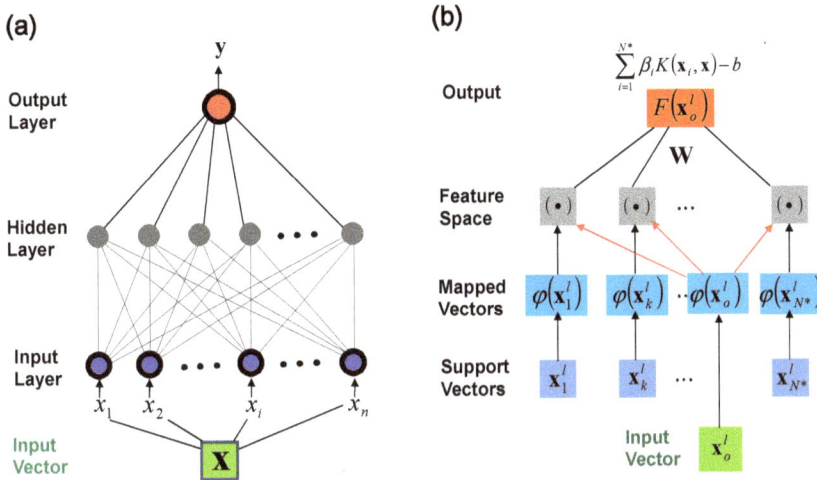

**Figure 5.** Schematic diagrams of the (**a**) artificial neural network (ANN) and (**b**) support vector machine (SVM) structure.

### 3.3. Time Series Modeling Strategy

In general, two types of strategies can be taken into account for time series modeling: direct and recursive prediction [33,34]. The direct prediction strategy always uses actual measured data as input components, thus, model accuracy is high, especially for short-term predictions. For long-term predictions, it requires separate models for every prediction horizon, which increases the computational burden and reduces the efficiency of the time series modeling. The direct prediction strategy can be expressed as follows:

$$\hat{y}_t^{DP} = M_h{}^{DP}\left(X_{\tau_i}^i, Y_{\tau_Y}\right), \quad \begin{cases} i = 1, \dots, n \\ X_{\tau_i}^i = x_{t-1}^i, \dots, x_{t-a_i}^i \\ Y_{\tau_Y} = y_{t-1}, \dots, y_{t-a_Y} \end{cases} \tag{10}$$

where $\hat{y}_t^{DP}$ is estimated target value at time t based on the direct prediction strategy, $M_h{}^{DP}$ is a time series model of the direct prediction for the prediction horizon of $h$; $X^i$ and $x^i$ are $i$-th exogenous input variable and its components, respectively; $Y$ is an autoregressive input variable that is identical to a target variable; $a_i$ and $a_Y$ are the number of past measurement data for $X^i$ and $Y$. The autoregressive input variable can be deleted if a model only uses exogenous inputs.

The recursive prediction strategy generally utilizes 1-lead time ahead of the direct prediction model repeatedly for estimating the autoregressive input components, which enables the model to perform a simulation and long-term prediction effectively. However, the error occurred from the direct prediction model in the previous time step can be accumulated continuously with time steps, which can deteriorate the model performance significantly [21,34]. Therefore, it is important to build an adequate direct prediction model for stable and accurate recursive prediction. The recursive prediction strategy is expressed as:

$$\hat{y}_t^{RP} = M_1^{DP}\left(X_{\tau_i}^i, \hat{Y}_{\tau_Y}\right), \quad \begin{cases} i = 1, \ldots, n \\ X_{\tau_i}^i = x_{t-1}^i, \ldots, x_{t-a_i}^i \\ \hat{Y}_{\tau_Y} = \hat{y}_{t-1}, \ldots, \hat{y}_{t-a_Y} \end{cases} \tag{11}$$

where $\hat{y}_t^{RP}$ is estimated target value at time $t$ based on the recursive prediction strategy, $M_1^{DP}$ is a 1-lead time direct prediction model and $\hat{Y}$ is the estimated autoregressive input variable. In this study, ANN- and SVM-based time series models with four types of input structures as combinations of tide level, rainfall, GWL, and FSL were designed for upper and lower FSL data, respectively. The number of past measurement data used for the component of each variable is described in Table 2. As an example, estimated upper FSL value at time $t$ based on the 1-lead time ahead of direct and recursive prediction strategies using T-R-G-F type model can be expressed as Equations (12) and (13), respectively.

$$\hat{y}_t^{DP} = M_1^{DP}\left(x_{t-4}^1, \ldots, x_{t-1}^1, x_{t-4}^2, \ldots, x_{t-1}^2, x_{t-3}^3, \ldots, x_{t-1}^3, \quad y_{t-4}, \ldots, y_{t-1}\right) \tag{12}$$

$$\hat{y}_t^{RP} = M_1^{DP}\left(x_{t-4}^1, \ldots, x_{t-1}^1, x_{t-4}^2, \ldots, x_{t-1}^2, x_{t-3}^3, \ldots, x_{t-1}^3, \quad \hat{y}_{t-4}, \ldots, \hat{y}_{t-1}\right) \tag{13}$$

where $x_{t-k}^1$, $x_{t-k}^2$, and $x_{t-k}^3$ are measured data of tide, rainfall, and GWL at time $t$-$k$, respectively; $y_t$ and $\hat{y}_t$ are measured and estimated FSL values at time $t$, respectively.

The data allocation for the model building and validation stages of the ANN and SVM models are described in Table 3.

Table 2. Model input structures and the number of components for each variable.

| Input Structures (Model Type) | | Number of Components for Variables | | | | |
|---|---|---|---|---|---|---|
| | | T | R | G | F | Total |
| Upper FSL | T-R | 4 | 4 | – | – | 8 |
| | T-R-F | 4 | 4 | – | 4 | 12 |
| | T-R-G | 4 | 4 | 3 | – | 15 |
| | T-R-G-F | 4 | 4 | 3 | 4 | 19 |
| Lower FSL | T-R | 8 | 4 | – | – | 12 |
| | T-R-F | 8 | 4 | – | 4 | 16 |
| | T-R-G | 8 | 4 | 5 | – | 17 |
| | T-R-G-F | 8 | 4 | 5 | 4 | 21 |

Notes: Where **T**: tide; **R**: rainfall; **G**: groundwater level; **F**: interface level.

Table 3. Data allocation for time series model building and validation.

| Data Type | | Data Allocation | |
|---|---|---|---|
| | | Model Building | Model Validation |
| Upper FSL | Num. data | 250 | 247 |
| | Time | 7:00 15 September–12:00 21 September | 13:00 21 September–23:00 5 October |
| Lower FSL | Num. data | 300 | 197 |
| | Time | 7:00 15 September–17:00 27 September | 18:00 27 September–23:00 5 October |

## 4. Results and Discussion

### 4.1. Direct Prediction of FSL

Time series models of 1-h direct prediction for the upper and lower FSL were constructed. The selected model parameters of the ANN and SVM for each type of input structure are described in Table 4.

**Table 4.** The selected ANN and SVM model parameters for the FSL prediction.

| Model Type | | ANN | | | SVM | | |
|---|---|---|---|---|---|---|---|
| | | HN | LR | MM | C | Eps | Sig |
| | T-R | 2 | 0.001 | 0.0 | 7.0 | 0.13 | 3.0 |
| Upper | T-R-F | 15 | 0.020 | 0.0 | 5.0 | 0.05 | 2.5 |
| FSL | T-R-G | 2 | 0.005 | 0.3 | 7.0 | 0.10 | 3.0 |
| | T-R-G-F | 5 | 0.005 | 0.0 | 10.0 | 0.05 | 3.0 |
| | T-R | 15 | 0.001 | 0.9 | 3.0 | 0.11 | 2.0 |
| Lower | T-R-F | 15 | 0.001 | 0.3 | 5.0 | 0.13 | 3.0 |
| FSL | T-R-G | 20 | 0.001 | 0.9 | 0.5 | 0.13 | 2.5 |
| | T-R-G-F | 10 | 0.001 | 0.0 | 5.0 | 0.13 | 3.0 |

Three performance criteria were used to evaluate the prediction ability of the ANN and SVM model, including the root mean squared error (RMSE), mean absolute relative error (MARE), and correlation coefficient (CORR), as follows:

$$RMSE = \sqrt{\frac{1}{N}\sum_{k=1}^{N}(\hat{y}_k - y_k)^2} \tag{14}$$

$$MARE = \frac{1}{N}\sum_{k=1}^{N}\frac{|\hat{y}_k - y_k|}{|y_k^{max} - y_k^{min}|} \times 100 \tag{15}$$

$$CORR = \frac{\sum_{k=1}^{N}(y_k - \bar{y})(\hat{y}_k - \bar{\hat{y}})}{\sqrt{\sum_{k=1}^{N}(y_k - \bar{y})^2}\sqrt{\sum_{k=1}^{N}(\hat{y}_k - \bar{\hat{y}})^2}} \tag{16}$$

where $y^{max}$ and $y^{min}$ denote the maximum and minimum values of the observed data, respectively; and $\bar{y}$ and $\bar{\hat{y}}$ denote the average observed and estimated values, respectively. The RMSE is a useful index for model performance evaluation when large errors are particularly undesirable as the errors are squared before they are averaged, which makes large errors have a relatively high weight. The MARE is the mean absolute error value divided by the range of observed data, thus it can compare the prediction results of the time series data showing different ranges of fluctuation. The CORR measures the extent and direction of a linear relationship between the observed and estimated values.

The model performance criteria of ANN and SVM models for 1-h direct prediction of the upper FSL show that overall prediction accuracy is high: RMSE was below 0.07 m, MARE below 10.82%, and CORR over 0.89 (Table 5). The model performance of the T-R-F and T-R-G-F type models (which uses past measurement data of FSL as input values) were better than that of the T-R and T-R-G type models. The T-R-G-F type SVM model showed the best performance for 1-h direct prediction of the upper FSL. The average value of each performance criteria shows that the overall direct prediction ability of the SVM was better than ANN for the upper FSL data in this study. The observed data and direct prediction results for the upper FSL are shown in Figure 6. Various types of ANN and

SVM models were trained adequately during the model building stage and there was no significant difference between the estimated values for the input structures.

**Table 5.** Model performance criteria values for the direct prediction of upper FSL.

| Model | Index | T-R | T-R-F | T-R-G | T-R-G-F | Average |
|-------|-------|-----|-------|-------|---------|---------|
| | RMSE (m) | 0.061 | 0.034 | 0.042 | 0.032 | 0.042 |
| ANN | MARE (%) | 10.329 | 5.901 | 7.083 | 5.613 | 7.232 |
| | CORR | 0.888 | 0.965 | 0.935 | 0.964 | 0.938 |
| | RMSE (m) | 0.072 | 0.029 | 0.038 | 0.023 | 0.040 |
| SVM | MARE (%) | 12.335 | 4.959 | 6.221 | 3.944 | 6.865 |
| | CORR | 0.882 | 0.982 | 0.954 | 0.980 | 0.949 |

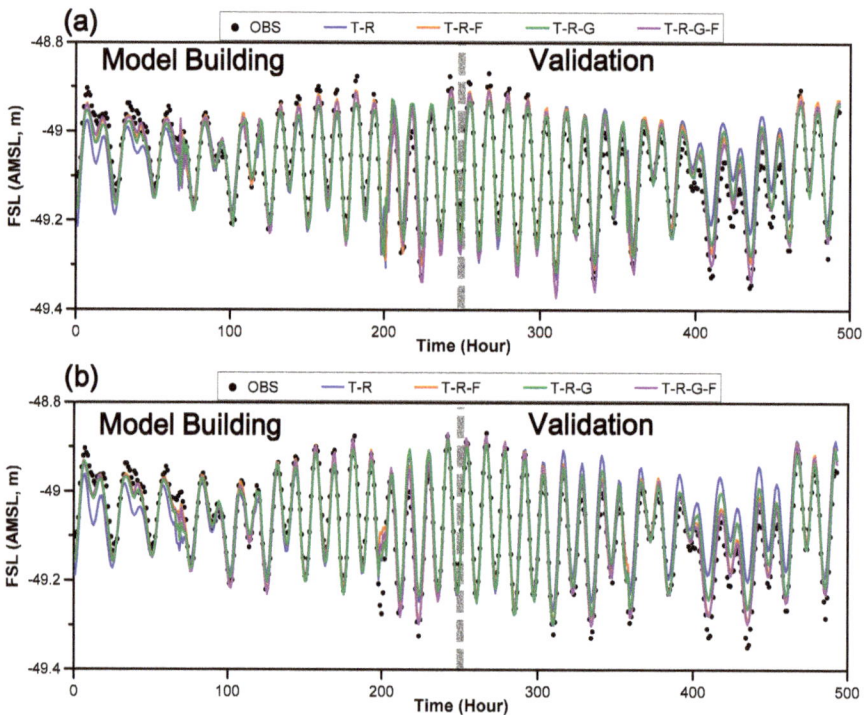

**Figure 6.** Direct prediction results for the upper FSL: (**a**) ANN and (**b**) SVM.

The quality performance of the direct prediction for the lower FSL was not as high as the upper FSL: the MARE values lay between 8.438% and 18.494%, and CORR values between 0.549 and 0.908 (Table 6). The RMSE values, ranging from 0.020 m to 0.040 m, were lower than the upper FSL prediction; however, this was not due to the prediction result for the lower FSL being better, but that the range of fluctuation of the lower FSL was smaller than the upper. The correlation of tide level and GWL with the lower FSL was weaker than the correlation with the upper FSL, which could cause deterioration in the model performance. The performance of the T-R-F and T-R-G-F type models was better than that of the T-R and T-R-G type models for lower FSL prediction, which was similar to the upper FSL prediction. The T-R-G-F type ANN model showed the best performance for 1-h direct prediction of the lower FSL; however, the average value of the performance criteria of the SVM models was better than the ANN. The observed data and direct prediction results for the lower FSL are shown in

Figure 7. The model building stage data included some abnormally high peaks (dashed circle) that probably had occurred due to pumping for an agricultural activity around the study site, which were not sufficiently trained by the ANN and SVM models, and can cause the underestimation at peak values in the validation stage.

**Table 6.** Model performance criteria values for the direct prediction of lower FSL.

| Model | Index | T-R | T-R-F | T-R-G | T-R-G-F | Average |
|-------|-------|-----|-------|-------|---------|---------|
| ANN | RMSE (m) | 0.034 | 0.020 | 0.040 | 0.020 | 0.028 |
| | MARE (%) | 15.314 | 8.438 | 18.494 | 8.623 | 12.717 |
| | CORR | 0.593 | 0.885 | 0.549 | 0.908 | 0.734 |
| SVM | RMSE (m) | 0.028 | 0.022 | 0.030 | 0.021 | 0.025 |
| | MARE (%) | 12.630 | 9.229 | 12.654 | 9.104 | 10.904 |
| | CORR | 0.777 | 0.859 | 0.733 | 0.867 | 0.809 |

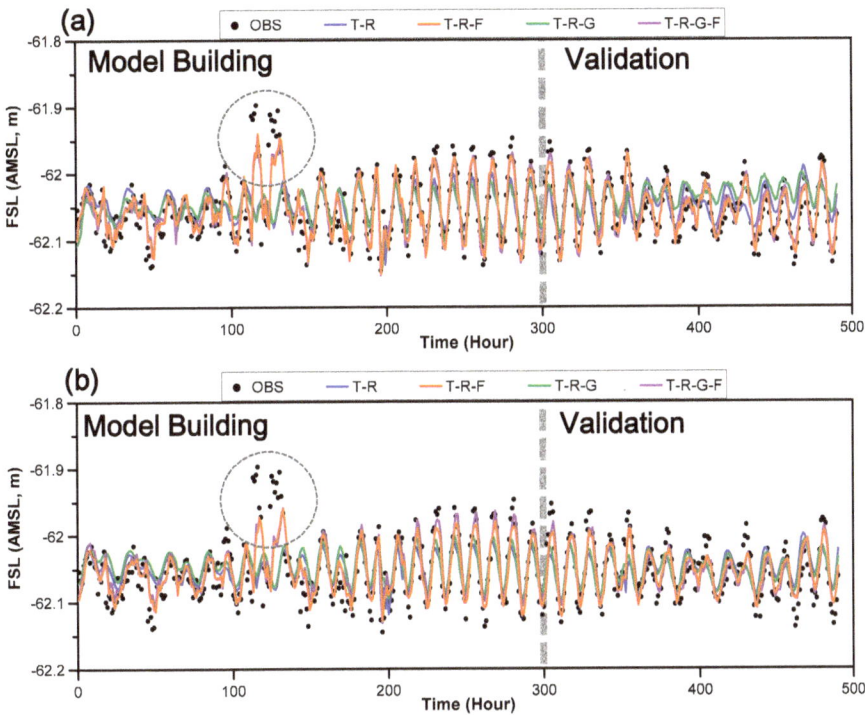

**Figure 7.** Direct prediction results for the lower FSL: (**a**) ANN and (**b**) SVM.

The results of the direct prediction showed that the best performance models for the upper and lower FSL were different and that the performance of the SVM model was less sensitive to the input structure of the ANN for the FSL data.

*4.2. Recursive Prediction of FSL*

The recursive prediction models of the upper and lower FSL were designed using the 1-h direct prediction models. The model performance criteria for the recursive prediction of the upper and lower FSL are described in Tables 7 and 8, respectively.

**Table 7.** Model performance criteria values for the recursive prediction of upper FSL.

| Model | Index | T-R | T-R-F | T-R-G | T-R-G-F | Average |
|-------|-------|-----|-------|-------|---------|---------|
| ANN | RMSE (m) | 0.061 | 0.061 | 0.042 | 0.056 | 0.055 |
|  | MARE (%) | 10.329 | 10.582 | 7.083 | 9.852 | 9.462 |
|  | CORR | 0.888 | 0.965 | 0.935 | 0.892 | 0.902 |
| SVM | RMSE (m) | 0.072 | 0.069 | 0.038 | 0.040 | 0.055 |
|  | MARE (%) | 12.335 | 12.255 | 6.221 | 7.043 | 9.463 |
|  | CORR | 0.882 | 0.920 | 0.954 | 0.943 | 0.925 |

**Table 8.** Model performance criteria values for the recursive prediction of lower FSL.

| Model | Index | T-R | T-R-F | T-R-G | T-R-G-F | Average |
|-------|-------|-----|-------|-------|---------|---------|
| ANN | RMSE (m) | 0.034 | 0.042 | 0.040 | 0.034 | 0.037 |
|  | MARE (%) | 15.314 | 16.837 | 18.494 | 14.937 | 16.395 |
|  | CORR | 0.593 | 0.420 | 0.549 | 0.806 | 0.592 |
| SVM | RMSE (m) | 0.028 | 0.034 | 0.030 | 0.035 | 0.032 |
|  | MARE (%) | 12.630 | 14.347 | 12.654 | 14.773 | 13.601 |
|  | CORR | 0.777 | 0.611 | 0.733 | 0.592 | 0.678 |

The overall model performance of the recursive prediction was lower than the direct prediction. The T-R-G type SVM models showed the best performance and the average values of the performance criteria of the SVM was superior to ANN for the recursive prediction of both the upper and lower FSL. The success of the recursive prediction highly relied on the generalization ability of the model to capture the relationship between the input and output variables of the given system as the observed data of the output variables are not available as input components. Based on the SRM, the inherent generalization ability of the SVM may capture the relationship between input and output data of this study more effectively than the ANN. The recursive prediction results of the ANN and SVM models for the upper and lower FSL are shown in Figures 8 and 9, respectively.

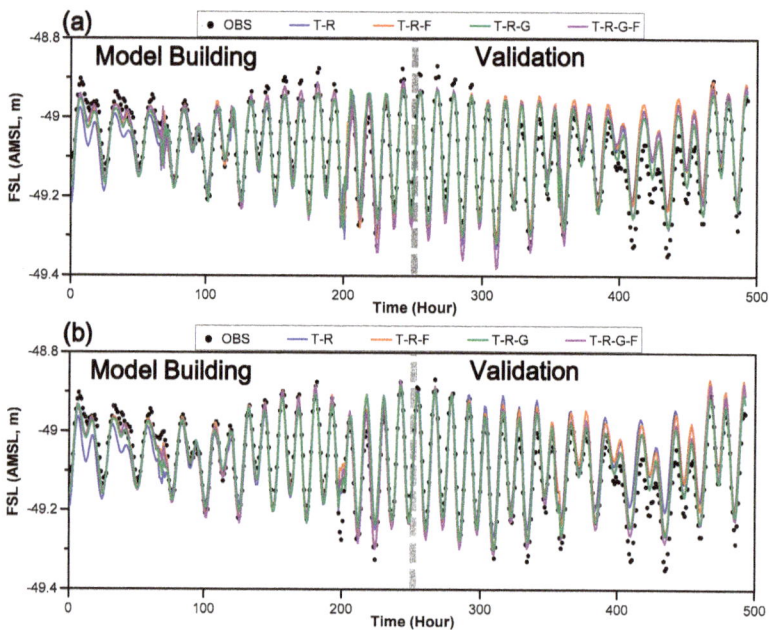

**Figure 8.** Recursive prediction results for the upper FSL: (a) ANN and (b) SVM.

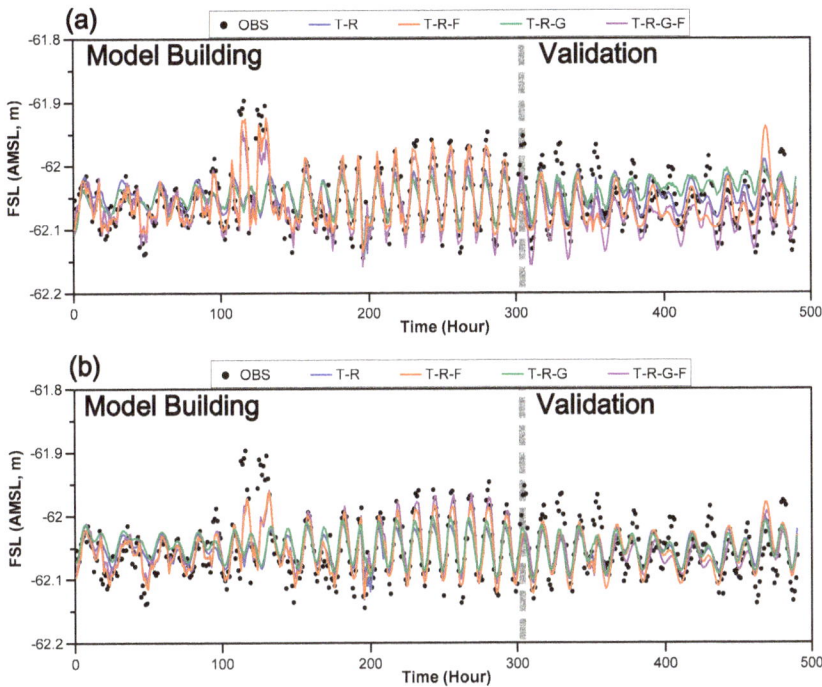

**Figure 9.** Recursive prediction results for the lower FSL: (**a**) ANN and (**b**) SVM.

The direct prediction strategy is efficient for the short-term prediction where a real-time measurement data of the target variable is available, and the recursive prediction strategy is necessary for the long-term prediction or the simulation of the target variable variation. However, as mentioned above and in Section 3.3, the error of the estimated target value can be accumulated with time steps in the recursive prediction strategy. Thus, it is important to build an adequate direct prediction model that learnt a response function of the given system. To evaluate the stability of the recursive model building, the RP-DP ratio [24] was calculated for T-R-F and T-R-G-F type models with 216 candidate model parameter groups:

$$RP - DP\ ratio = \frac{RMSE\ of\ the\ recursive\ prediction}{RMSE\ of\ the\ direct\ prediction} \tag{17}$$

The RP-DP ratio value stands for the extent of the consistency between the direct and recursive prediction models. Thus, a narrower distribution with lower values of the RP-DP ratio indicates a higher possibility that a recursive prediction model of high consistency with a direct prediction model is selected. The calculated RP-DP ratio values of the ANN models were more distributed than the SVM for both the T-R-F and T-R-G-F type models and the upper and lower FSL (Figure 10). These results indicate that the SVM method is more efficient and stable than the ANN for the recursive prediction of the FSL data in this study.

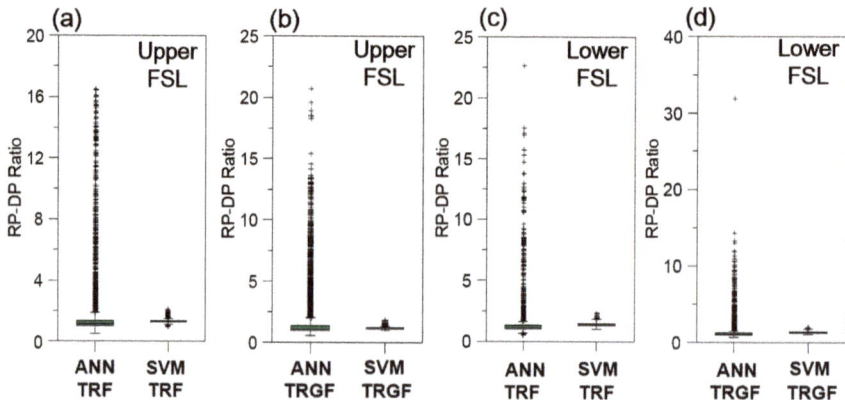

**Figure 10.** Comparison of RP-DP ratio values for ANN and SVM models: (**a**) T-R-F type model for upper FSL; (**b**) T-R-G-F type model for upper FSL; (**c**) T-R-F model for lower FSL; (**d**) T-R-G-F type model for lower FSL.

## 5. Summary and Conclusions

In this study, the temporal variation of the upper and lower FSL was monitored using interface eggs at the HD2 observatory on Jeju Island, South Korea. The ANN- and SVM-based time series models of FSL prediction were developed and their performance compared. The result of the direct prediction shows that the T-R-G-F type ANN model was best for upper FSL prediction and the T-R-G-F type SVM model for the lower FSL. The T-R-G type SVM model was best for the recursive prediction of both upper and lower FSL. The average values of the model performance criteria indicated that the overall prediction ability of the SVM model was superior to the ANN. The analysis of the RP-DP ratio distribution showed that the SVM-based recursive prediction model was more stable and efficient than the ANN for FSL prediction of the study site.

The monitoring and prediction of FSL is necessary for the sustainable use of groundwater resources in coastal aquifers. The groundwater is the sole and main water source of Jeju Island and the local government has installed and operated a saltwater intrusion monitoring network. It is expected that the developed model for FSL prediction can be a useful tool in the future management of groundwater resources in coastal areas.

**Acknowledgments:** This research was supported by Korea Ministry of Environment as "GAIA (Geo-Advanced Innovative Action) Project (#2016000530004)".

**Author Contributions:** Heesung Yoon and Kyoochul Ha designed and performed numerical experiments and wrote the paper; Yongcheol Kim and Soo-Hyung Lee set the monitoring system and conducted the time series analysis; Gee-Pyo Kim measured the vertical profile of the electrical conductivity at the study site.

**Conflicts of Interest:** The authors declare no conflicts of interest.

## References

1. Kim, K.; Seong, H.; Kim, T.; Park, K.; Woo, N.; Park, Y.; Koh, G.; Park, W. Tidal effects on variations of fresh-saltwater interface and groundwater flow in a multilayered coastal aquifer on a volcanic island (Jeju Island, Korea). *J. Hydrol.* **2006**, *330*, 525–542. [CrossRef]
2. Liu, C.K.; Dai, J.J. Seawater intrusion and sustainable yield of basal aquifers. *J. Am. Water Resour. Assoc.* **2012**, *48*, 861–870. [CrossRef]
3. Kim, Y.; Yoon, H.; Kim, K.P. Development of a novel method to monitor the temporal change in the location of the freshwater-saltwater interface and time series models for the prediction of the interface. *Environ. Earth Sci.* **2016**, *75*, 882–891. [CrossRef]

4.  Gingerich, S.B.; Voss, C.I. Thee-dimensional variable-density flow simulation of a coastal aquifer in southern Oahu, Hawaii, USA. *Hydrogeol. J.* **2005**, *13*, 436–450. [CrossRef]

5.  Wener, A.D.; Gallagher, M.R. Characterization of sea-water intrusion in the Pioneer Valley, Australia using hydrochemistry and three-dimensional numerical modelling. *Hydrogeol. J.* **2006**, *14*, 1452–1469. [CrossRef]

6.  Guo, W.; Langevin, C.D. *User's Guide to SEWAT: A Computer Program for Simulation of Three-Dimensional Variable-Density Ground-Water Flow*; United States Geological Survey: Tallahassee, FL, USA, 2002; p. 73.

7.  Rozell, D.J.; Wong, T. Effects of climate change on groundwater resources at Shelter Island, New York State, USA. *Hydrogeol. J.* **2010**, *18*, 1657–1665. [CrossRef]

8.  Yechieli, Y.; Shalev, E.; Wollman, S.; Kiro, Y.; Kafri, U. Response of the Mediterranean and Dead Sea coastal aquifers to sea level variations. *Water Resour. Res.* **2010**, *46*, W12550. [CrossRef]

9.  Bardossy, A. Calibration of hydrological model parameters for ungauged catchments. *Hydrol. Earth Syst. Sci.* **2007**, *11*, 703–710. [CrossRef]

10. Pollacco, J.A.P.; Ugalde, J.M.S.; Angulo-Jaramillo, R.; Braud, I.; Saugier, B. A linking test to reduce the number of hydraulic parameters necessary to simulate groundwater recharge in unsaturated zone. *Adv. Water Resour.* **2008**, *31*, 355–369. [CrossRef]

11. Zealand, C.M.; Burn, D.H.; Simonovic, S.P. Short-term streamflow forecasting using artificial neural networks. *J. Hydrol.* **1999**, *214*, 32–48. [CrossRef]

12. Akhtar, M.K.; Corzo, G.A.; van Andel, S.J.; Jonoski, A. River flow forecasting with artificial neural networks using satellite observed precipitation pre-processed with flow length and travel time information: Case study of the Ganges river basin. *Hydrol. Earth Syst. Sci.* **2009**, *13*, 1607–1618. [CrossRef]

13. Hu, T.S.; Lam, K.C.; Ng, S.T. A modified neural network for improving river flow prediction. *Hydrol. Sci. J.* **2005**, *50*, 299–318. [CrossRef]

14. Coulibaly, P.; Anctil, F.; Aravena, R.; Bobee, B. Artificial neural network modeling of water table depth fluctuations. *Water Resour. Res.* **2001**, *37*, 885–896. [CrossRef]

15. Coppola, E.; Rana, A.J.; Poulton, M.M.; Szidarovszky, F.; Uhl, V.V. A neural network model for predicting aquifer water level elevations. *Groundwater* **2005**, *43*, 231–241. [CrossRef] [PubMed]

16. Mohanty, S.; Jha, M.K.; Kumar, A.; Sudheer, K.P. Artificial neural network modeling for groundwater level forecasting in a river island of eastern India. *Water Resour. Manag.* **2010**, *24*, 1845–1865. [CrossRef]

17. Liong, S.Y.; Sivapragasam, C. Flood stage forecasting with support vector machines. *J. Am. Water Resour. Assoc.* **2002**, *38*, 173–186. [CrossRef]

18. Yu, P.S.; Chen, S.T.; Chang, I.F. Support vector regression for real-time flood stage forecasting. *J. Hydrol.* **2006**, *328*, 704–716. [CrossRef]

19. Asefa, T.; Kemblowski, M.W.; Urroz, G.; McKee, M.; Khalil, A. Support vectors-based groundwater head observation networks design. *Water Resour. Res.* **2004**, *40*, W1150901. [CrossRef]

20. Gill, M.K.; Asefa, T.; Kaheil, Y.; McKee, M. Effect of missing data on performance of learning algorithms for hydrologic prediction: Implication to an imputation technique. *Water Resour. Res.* **2007**, *43*, W07416. [CrossRef]

21. Yoon, H.; Hyun, Y.; Ha, K.; Lee, K.K.; Kim, G.B. A method to improve the stability and accuracy of ANN- and SVM-based time series models for long-term groundwater level predictions. *Comput. Geosci.* **2016**, *90*, 144–155. [CrossRef]

22. Krishna, B.; Satyaji Rao, Y.R.; Vijaya, T. Modelling groundwater levels in an urban coastal aquifer using artificial neural networks. *Hydrol. Process.* **2008**, *22*, 1180–1188. [CrossRef]

23. Nayak, P.C.; Satyaji Rao, Y.R.; Sudheer, K.P. Groundwater level forecasting in a shallow aquifer using artificial neural network approach. *Water Resour. Manag.* **2006**, *20*, 77–90. [CrossRef]

24. Yoon, H.; Jun, S.-C.; Hyun, Y.; Bae, G.-O.; Lee, K.-K. A comparative study of artificial neural networks and support vector machines for predicting groundwater levels in a coastal aquifer. *J. Hydrol.* **2011**, *396*, 128–138. [CrossRef]

25. Yang, X.; Zhang, H.; Zhou, H. A hybrid methodology for salinity time series forecasting based on wavelet transform and NARX neural networks. *Arab. J. Sci. Eng.* **2014**, *39*, 6895–6905. [CrossRef]

26. Wan, Y.; Wan, C.; Hedgepeth, M. Elucidating multidecadal saltwater intrusion vegetation dynamics in a coastal floodplain with artificial neural networks and aerial photography. *Ecohydology* **2015**, *8*, 309–324. [CrossRef]

27. ASCE Task Committee on Application of Artificial Neural Networks in Hydrology. Artificial neural networks in hydrology I: Preliminary concepts. *J. Hydrol. Eng.* **2000**, *5*, 115–123.

28. Maier, H.R.; Dandy, G.C. Neural networks for the prediction and forecasting of water resources variables: A review of modeling issues and applications. *Environ. Model. Softw.* **2000**, *15*, 101–124. [CrossRef]

29. Rumelhart, D.E.; McClelland, J.L. *The PDP Research Group, Parallel Distributed Processing: Explorations in the Microstructure of Cognition*; MIT Press: Cambridge, MA, USA, 1986.

30. Vapnik, V.N. *The Nature of Statistical Learning Theory*; Springer: New York, NY, USA, 1995.

31. Platt, J.C. Fast training of support vector machines using sequential minimal optimization. In *Advances in Kernel Methods-Support Vector Learning*, 2nd ed.; Schölkopf, B., Burges, C., Smola, A., Eds.; MIT Press: Cambridge, MA, USA, 1998.

32. Scholkopf, B.; Smola, A.J. *Learning with Kernels: Support Vector Machines, Regularization, Optimization, and Beyond*; MIT Press: Cambridge, MA, USA, 2002.

33. Ji, Y.; Hao, J.; Reyhani, N.; Lendasse, A. Direct and recursive prediction of time series using mutual information selection. In Proceedings of the International Work-Conference on Artificial Neural Networks, Barcelona, Spain, 8–10 June 2005; Volume 3512, pp. 1010–1017.

34. Herrera, L.J.; Pomares, H.; Rojas, I.; Guillen, A.; Prieto, A.; Valenzuela, O. Recursive prediction for long term time series forecasting using advanced models. *Neurocomputing* **2007**, *70*, 2870–2880. [CrossRef]

*water*

MDPI

*Article*

# Sharp Interface Approach for Regional and Well Scale Modeling of Small Island Freshwater Lens: Tongatapu Island

**Roshina Babu [1], Namsik Park [1,\*], Sunkwon Yoon [2] and Taaniela Kula [3]**

[1] Department of Civil Engineering, Dong-A University, Busan 49315, Korea; roshinababu@gmail.com
[2] Integrative Climate Research Team, APEC Climate Center, Busan 48058, Korea; skyoon@apcc21.org
[3] Ministry of Lands & Natural Resources, Nuku'alofa, Tonga; taanielakula@gmail.com
\* Correspondence: nspark@dau.ac.kr; Tel: +82-51-200-7629

Received: 28 September 2018; Accepted: 9 November 2018; Published: 12 November 2018

**Abstract:** Sustainable management of small island freshwater resources requires an understanding of the extent of freshwater lens and local effects of pumping. In this study, a methodology based on a sharp interface approach is developed for regional and well scale modeling of island freshwater lens. A quasi-three-dimensional finite element model is calibrated with freshwater thickness where the interface is matched to the lower limit of the freshwater lens. Tongatapu Island serves as a case study where saltwater intrusion and well salinization for the current state and six long-term stress scenarios of reduced recharge and increased groundwater pumping are predicted. Though no wells are salinized currently, more than 50% of public wells are salinized for 40% decreased recharge or increased groundwater pumping at 8% of average annual recharge. Risk of salinization for each well depends on the distance from the center of the well field and distance from the lagoon. Saltwater intrusions could occur at less than 50% of the previous estimates of sustainable groundwater pumping where local pumping was not considered. This study demonstrates the application of a sharp interface groundwater model for real-world small islands when dispersion models are challenging to be implemented due to insufficient data or computational resources.

**Keywords:** sharp interface numerical modeling; freshwater lens; saltwater intrusion; well salinization; small islands; Tongatapu

## 1. Introduction

Groundwater occurring in the form of a thin lens floating on denser seawater is the primary source of freshwater in most small islands [1]. This limited groundwater resource is highly vulnerable to saltwater intrusion due to natural causes such as droughts, storm surges, sea level rise, and anthropogenic activities such as increased groundwater withdrawals [2,3]. Many small islands in the Pacific face groundwater shortages during droughts associated with El Niño—Southern Oscillation (ENSO) events [2] which are predicted to be frequent in the future due to global warming [4]. Public water supply in small raised coral islands of the Pacific including Tongatapu and Niue [5] relies on multiple vertical wells. Large-scale localized and unplanned abstraction can expedite the salinization of the freshwater lens [2]. Over-pumping can severely contaminate the aquifer, and hence, it is essential to modify and manage the pumping rates, especially during droughts. Management of groundwater withdrawal requires knowledge about the extent of freshwater lens and identification of wells that are under high risk of saltwater intrusion [6]. Hence, regional and wellscale modeling of freshwater lens are essential for planning water management strategies in small islands.

Various numerical models are available to assess the freshwater resources in small islands [7]. However, quantitative assessment of freshwater resources and seawater intrusion impacts have been performed in only a small number of real-world islands [8,9]. Numerical models, which consider the dispersion zone between the freshwater lens and underlying seawater, can closely represent typical island groundwater processes [10–12]. With fine discretization of space and time and sufficient monitoring data, dispersion models can provide a quantitative assessment of freshwater lens including the local effects of pumping [7,11,13]. Proper monitoring of well salinities is often scarce, if not absent, in many real islands [2] and it is difficult to obtain a sufficient quantity and quality of monitoring data for calibration [14,15]. Hence, due to the high computational demands and lack of sufficient monitoring data for calibration of heads and salinity, field-scale modeling of regional and well scale freshwater lens have been rarely attempted using dispersion models [7,11], and use of empirical models are suggested for well scale modeling [16]. As a result, use of a single numerical model to evaluate both the island-wide extent of freshwater lens and the local effects of pumping remains a challenge when the size of an island is not so small.

Numerical models that assume freshwater and saltwater to be immiscible and separated by a sharp interface have been used for the evaluation of risks of saltwater intrusion to freshwater resources when dispersion models are difficult to be implemented in regional scales [17]. Sharp interface numerical models have lesser data and computational requirements and have been used to predict salinity at pumping wells [18,19]. Hence, a sharp interface approach is used in this study for regional and well scale modeling of the freshwater lens on an island. It ensures a single numerical model is capable of providing a first-hand prediction of the extent of freshwater lens and the local effects of pumping.

Sharp interface two-phase numerical modeling was applied for Nantucket Island [20] to assess the impact of projected pumping on the freshwater lens. The model predicted the general possibility of saltwater upconing into the well field when the interface was within the intake of deepest well, though no attempt was made to quantify individual well salinization. Freshwater heads were used for calibration of the model, and no island-wide comparison of the model predicted freshwater thickness with the field observations were done. The midpoint of freshwater-saltwater mixing zone was matched to the sharp interface and hence the encroachment of water with salinity higher than the freshwater limit (chloride content of 250 mg/L) into well screen could not be identified.

When a numerical model for groundwater flow is developed, the model is generally calibrated with observed groundwater heads [21]. Some issues associated with the monitoring data of groundwater heads in small islands make it practically difficult to be used for calibration. The highest freshwater heads in small islands are typically of the order of 0.2 to 0.5 m above the mean sea level [2,22] due to the high hydraulic conductivity of the aquifer. Since the depth of the interface estimated by Ghyben-Herzberg's approximation [23,24] is 40 times freshwater heads, highly accurate groundwater heads in the order of centimeter accuracy may be required to predict freshwater thickness. Additionally, due to the insufficiency of salinity monitoring data [2,7], the vertical profile of groundwater density is ambiguous, and so conversion of observed heads to equivalent freshwater heads is also challenging. In some islands like Tongatapu, which is the case study selected here, accurate land surface elevations are not available at monitoring locations.

Freshwater thickness is the most critical parameter to be assessed for groundwater management in small islands [7]. In this study, the freshwater thickness is selected for calibration of the groundwater model to overcome the challenges associated with the use of freshwater heads in calibration. The two-fluid sharp interface approach used in this study [25] represents groundwater flow in terms of freshwater and saltwater heads. Though it is unable to predict the nature of the transition zone, it can predict the response of the interface [26] and pumping wells [18,27] to applied stresses. By adopting the upper limit of freshwater electrical conductivity as the interface, it is possible to predict the extent of freshwater thickness in the island.

The objective of this work is to develop a methodology based on a sharp interface numerical model to conduct regional and well scale modeling of island freshwater lens under long-term stresses. The methodology is applied to Tongatapu, a small Pacific Island. A numerical model is developed and calibrated for the average recharge and the current pumping condition and is then used to assess the effect of long-term stresses of recharge and pumping variations on saltwater intrusion and well salinization. This work expands the capacity of the sharp interface numerical model to regional and well scale modeling of freshwater lens in real-world small islands. The methodology developed can be applied to other similar island systems for the prediction of freshwater lens and to quantify the local effects of pumping. The impact of long-term stresses on saltwater intrusion along with well salinization is evaluated using the groundwater system of Tongatapu and generalized outcomes are relevant to other similar island systems with concentrated pumping. Tongatapu is the largest and the most populated island of the Kingdom of Tonga, a Pacific island nation. There has been an increasing trend of migration to Tongatapu, more specifically, to its capital Nuku'alofa from outer islands [28]. Groundwater extracted from vertical wells meets the public water supply demands for Nuku'alofa due to the lack of surface water sources [29]. With the increase in groundwater demands and expected climate changes in the future, there has been an increase in public concern over groundwater availability. Previous numerical modeling studies on Tongatapu [30,31] used a large grid size (about 1 km) for simulations and ignored the effects of groundwater pumping and future change in recharge patterns. For small islands like Tongatapu where groundwater pumping is concentrated over a small area, the risk of well salinization is high due to saltwater upconing [2]. The results of this study can help to identify the wells with a high risk of salinization, which was rarely attempted in earlier modeling studies and can contribute to planning and management of groundwater development.

## 2. Materials and Methods

### 2.1. Study Area

Tongatapu ($21°03'$–$21°16'$ S and $175°02'$–$175°21'$ W) is the main island of Kingdom of Tonga (Figure 1). It has a total area of 259 km$^2$ and is the location of the national capital Nuku'alofa. The population of Tongatapu was 75,416 (73% of total population of the Kingdom of Tonga), with a density of 290 persons per square kilometers in 2011 [32]. The island is 31.5 km along its Northwest-Southeast directions and 18.5 km along the Northeast-Southwest direction.

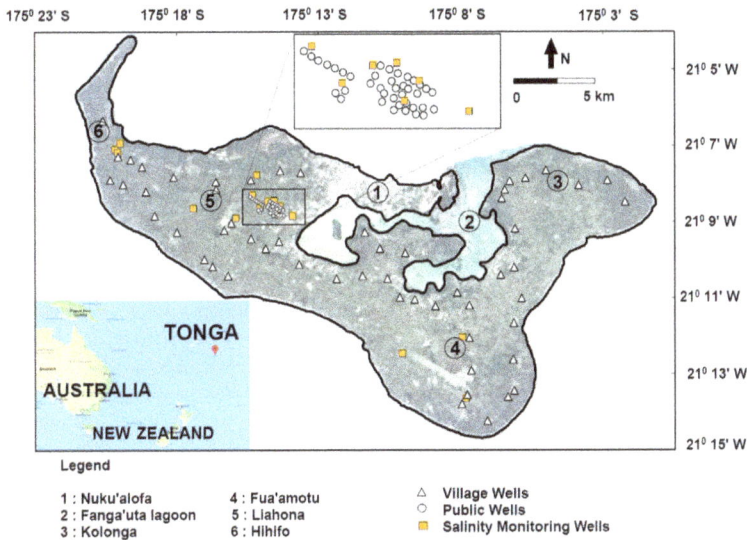

**Figure 1.** Island of Tongatapu.

Tongatapu is a tilted raised limestone island with a maximum elevation of 65 m [22] above mean sea level. The aquifer consists of karst-type limestone that is generally very porous and has hydraulic conductivity ranging from 500 to 3600 m/day [22,33]. The drill logs indicated a single layer of limestone aquifer [34]. The freshwater stored in the limestone aquifer is generally in the form of a lens floating on denser salt water. The water table of the island of Tongatapu is almost flat with an average hydraulic head of 0.4 m [22], and the groundwater lens with an electrical conductivity less than 2500 µS/cm was measured to have a maximum thickness of about 14 m [22,33,35]. The freshwater thickness at a few locations like Liahona and Fua'amotu (Figure 1) was estimated from monitoring wells. However, it is less understood across the remainder of Tongatapu as only limited observation data is available [34]. Fanga'uta lagoon (Figure 1) opens to the sea and salinity is reported to be approximately 25,700 to 32,900 ppm dissolved solids [36].

The semi-tropical climate of Tonga is highly vulnerable to the effects of El Nino, which frequently coincides with drought incidence [37]. Tongatapu has a mean temperature of about 24 °C [38]. Mean annual rainfall for Nuku'alofa, Tongatapu from the year 1993 to 2016 was 1781 mm. The rainfall pattern is characterized by the contrast between a wet season (November–April) which accounts for over 60% of total annual rainfall and a dry season (May–October). Values ranging from 20–35% of the mean annual rainfall have been reported as annual recharge rates in Tongatapu [22,33,36].

Groundwater extraction in Tongatapu is divided into public/urban and village/rural water supply. The public well field at Mataki'eua-Tongamai (Figure 1) is the main source of water supply to the capital city Nuku'alofa. The well field consists of 39 vertical wells concentrated over an area of 0.57 km$^2$ [29]. As per the Ministry of Lands and Natural Resources (MLNR) Tonga, the groundwater extraction from the public well field in 2016 was 10410 m$^3$/day. Rural water supply varies between villages and most of the wells are located at the edges of villages. It is estimated that there are about 200 village wells in Tongatapu [36]. It is observed that the current pumping from village wells is difficult to be estimated, as there are no flow meters installed [39].

There are sixteen salinity monitoring boreholes (SMB) throughout the island, ten around the Mataki'eua-Tongamai well field, and three each around Hihifo and Fua'amotu (Figure 1). Each salinity monitoring borehole (SMB) in Tongatapu consists of a nested group of four to six measuring tubes with 1 m long screen near the bottom of each tube (Figure 2). A screen of each tube either contains fresh or non-fresh water. The freshwater limit adopted in Tongatapu is based on the European Commission (1998) maximum electrical conductivity (EC) of 2500 µS/cm [22]. Figure 2 illustrates a situation where the screens of tubes 3 and 4 contain freshwater, while those of 1 and 2 contain non-freshwater. In this case, the lower boundary of the freshwater lens lies between screens 2 and 3. The freshwater thickness at the monitoring location is estimated as depth to EC value of 2500 µS/cm from the water table. The Tonga government (MLNR, Tonga 2017) adopted linear interpolation between EC values measured below and above 2500 µS/cm and their corresponding depths to obtain the freshwater thickness. This linear interpolation method results in a slightly conservative estimate of freshwater thickness compared to that based on relative salinity used for the estimation of freshwater thickness based on monitoring data in Majuro Atoll [40].

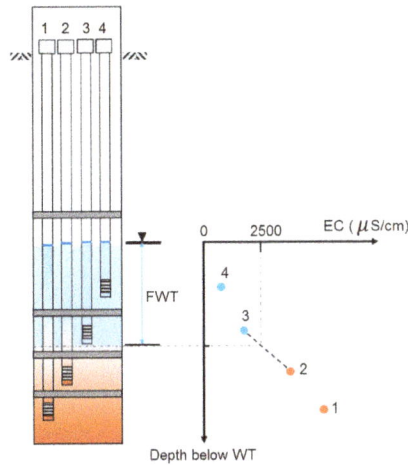

**Figure 2.** Construction of a Salinity Monitoring Borehole (SMB) (not to scale) and determination of freshwater thickness (FWT).

*2.2. Sharp Interface Freshwater-Seawater Groundwater Flow Numerical Model*

2.2.1. Governing Equations

In this study, a sharp-interface numerical model described by Huyakorn et al. [25] is chosen to simulate fresh and saline groundwater in the island. The sharp interface model is less rigorous than a dispersion model in treating physics relevant to the problem at hand. However, the former model requires fewer data and low-resolution grids so that it can be applied with relative ease to a large area. This model considers the flow dynamics of both freshwater and saltwater. The sharp interface freshwater-seawater model for groundwater flow with free surface consists of two vertically integrated governing equations, one for freshwater flow and the other for saltwater flow:

$$\frac{\partial}{\partial x}\left(K_x^f b^f \frac{\partial h^f}{\partial x}\right) + \frac{\partial}{\partial y}\left(K_y^f b^f \frac{\partial h^f}{\partial y}\right) = S_y^f \frac{\partial h^f}{\partial t} - \theta \frac{\partial \zeta}{\partial t} - Q^f \tag{1}$$

$$\frac{\partial}{\partial x}\left(K_x^s b^s \frac{\partial h^s}{\partial x}\right) + \frac{\partial}{\partial y}\left(K_y^s b^s \frac{\partial h^s}{\partial y}\right) = S_y^s \frac{\partial h^s}{\partial t} + \theta \frac{\partial \zeta}{\partial t} - Q^s \tag{2}$$

where the superscripts f and s refer to freshwater and saltwater, respectively, and $h$ is the hydraulic head, which is vertically averaged for the freshwater zone and saltwater zone. $b^f$ and $b^s$ are the thickness of freshwater and saltwater zones in the aquifer. $K_x^f$, $K_y^f$ and $K_x^s$, $K_y^s$ are the hydraulic conductivities in x and y directions for freshwater and saltwater, respectively. $\theta$ is effective porosity, $\zeta$ is the height of saltwater- freshwater interface above the datum. $S_y^f$ and $S_y^s$ are specific yield of the aquifer in freshwater and saltwater zone. The coupled freshwater and saltwater flow equations are solved simultaneously to obtain freshwater heads and saltwater heads. The interface elevation is calculated by equating pressures at the interface [41];

$$\zeta = \frac{\rho_f}{(\rho_s - \rho_f)}\left[(\rho_s/\rho_f)h^s - h^f\right] \tag{3}$$

where $\rho_f$ and $\rho_s$ are the freshwater and saltwater densities.

When the freshwater-saltwater interface is within the well screen, it is assumed that both freshwater and saltwater are extracted simultaneously. The total pumping rate $Q^T$ at the well is considered to be the sum of $Q^f$ and $Q^s$, the volumetric fluxes per unit volume of aquifer due to pumping of freshwater and saltwater, respectively. $Q^f$ and $Q^s$ are related to the total pumping rate $Q^T$ as follows.

$$Q^f = \left( \frac{K^f L^f}{K^f L^f + K^s L^s} \right) Q^T \tag{4}$$

$$Q^s = \left( \frac{K^s L^s}{K^f L^f + K^s L^s} \right) Q^T \tag{5}$$

where $L$ is the total length of the screen, $L^f$ and $L^s$ are the length of the screen exposed to freshwater and saltwater, respectively. $K^f L^f$ and $K^s L^s$ represent transmissivities of freshwater and saltwater regions, respectively. The saltwater ratio (SWR) at a pumping well is obtained as the ratio of saltwater pumped to the total pumping at the well.

$$SWR = \left( \frac{Q^s}{Q^f + Q^s} \right) \tag{6}$$

Despite the simplicity of the above approach, it was demonstrated that saltwater ratios at pumping wells could be modeled with reasonable accuracy [18,27].

A modified Galerkin finite element method was used to approximate governing equations [25]. The numerical model [25] has been improved in various aspects and applied to diverse groundwater problems such as assessment of saltwater intrusion due to groundwater developments [42], evaluations of potential groundwater development [43], coastal groundwater discharge for complex coastlines [44], estimation of saltwater contents in pumping wells [18,27], assessment of impacts of combined fresh and saltwater developments [45], and evaluation of the effectiveness of freshwater injection and/or saltwater pumping to prevent saltwater intrusions [46,47].

### 2.2.2. Conceptual Model for Groundwater Flow in Tongatapu Island

The numerical model was applied to the entire Tongatapu Island. Rectangular finite elements of 100 m × 100 m are used for areas with less groundwater development, and elements of 25 m × 25 m are applied for the central part of the island where public wells are concentrated. The grid was selected based on a grid convergence test. Consequently, the numerical model consists of 89,200 elements and 89,824 nodes. A single homogeneous isotropic unconfined aquifer with the base elevation at −40 m with reference to mean sea level (MSL) is considered. The effective porosity of the aquifer is assumed to be uniform as 0.3 [22]. Densities of freshwater and seawater are 1000 and 1024 kg/m³, respectively. A no-flow boundary is assumed at the bottom surface of the aquifer. At those nodes along the coastline and lagoon first type and efflux-only third type boundary conditions are specified for the saltwater equation and the freshwater equation, respectively. Recharge is applied with flux boundary conditions at all internal nodes.

Recharge was estimated from a water balance model provided by APEC Climate Center (APCC) based on the soil water balance model for a small island with no runoff [22]. The surface runoff was neglected as in other small islands due to flat terrain, high infiltration, and lack of surface flow observations. The water balance model parameters used in the present study (Table S1) are based on the sensitivity analysis by White et al. [22] for Tongatapu. As the daily rainfall data was unavailable, monthly rainfall data was used as input. The available monthly rainfall from 1993–2016 at Nuku'alofa was used to estimate the average annual recharge as 570 mm/year. This is 32% of the average annual rainfall during the period and falls within the range of recharge estimates by previous studies [22,33,36].

The locations of 39 public wells and 56 village wells are known [22], and hence a total of 95 pumping wells have been considered (Figure 1). Actual depths, lengths of well screens and individual pumping rates are not known. Screens are assumed to extend from −2 m to −5 m (MSL) for public wells and 0 m to −1 m (MSL) for village wells [34]. Groundwater extraction estimates for the public well field in 2016 was 10,410 m³/day (MLNR Tonga, 2017). The above pumping rate is assumed to be equally distributed in the 39 public wells as 267 m³/day/well. Tonga Water Board estimates rural per capita demand to be 80 L per person per day with rainwater availability [48].

Using the rural population of Tongatapu as per 2011 census and assuming 50% losses [22] the rural groundwater extraction is estimated to be 6300 m$^3$/day in this study. This is equally distributed to the 56 village wells included in the model as 112.5 m$^3$/day/well. Thus, the total groundwater extraction for Tongatapu is approximated to be 16,710 m$^3$/day. This value is close to estimates of groundwater demands for Tongatapu by Waterloo and Ijzermans [39].

*2.3. Calibration of the Model*

Depths to the groundwater table and electrical conductivity (EC) values are measured at salinity monitoring boreholes (SMB) in Tongatapu. However, measured depths cannot be converted to groundwater heads as land surface elevations of the reference point at SMBs from where depth to water table is measured and is not known precisely. In this study, freshwater thickness values are used as the calibration targets. The aquifer parameter of horizontal hydraulic conductivity based on available data was selected and later adjusted to obtain simulated freshwater thickness, which satisfactorily matches observed freshwater thickness in 16 salinity monitoring boreholes (SMB) distributed across the island (Figure 1). The Tonga government (MLNR Tonga) provided an average freshwater thickness in each SMB based on quarterly salinity monitoring data from July 1997 to April 2017. Seven out of the sixteen salinity monitoring boreholes were constructed in 1997, and the rest were added in 2012. On average, 56 observations are available for the older SMBs and 20 for the new SMBs. The average coefficient of variations (standard deviation divided by mean) are 0.18 and 0.13 for old and new SMBs, respectively, indicating a relatively stable freshwater thickness at each SMB. The manual trial and error model calibration, though unable to lead to a unique solution, is constrained based on the data available [26]. The simulated groundwater flow field was based on the steady state assumption despite temporal variation in recharge. Groundwater pumping was known to begin in 1960's and has been increasing since then. Nevertheless, it is assumed that the present groundwater flow field has reached the equilibrium state to the current pumping rate.

*2.4. Long-Term Stresses*

2.4.1. Impact of the Current Pumping

Simulations start with an initial condition of the aquifer fully saturated with saltwater below the MSL. Transient simulations are continued till the total saltwater and freshwater volume in the domain reached the steady state. Pumping wells are removed from the calibrated model to estimate the groundwater flow for the predevelopment condition prior to the groundwater development. All other conditions except for the pumping wells were assumed applicable to the predevelopment simulation. The predevelopment model was also run under the steady state assumption.

2.4.2. Impact of Long-Term Stresses

In this study, two major long-term stresses, droughts, and overpumping, which can cause saltwater intrusion and well salinization in Tongatapu are considered. Hydrological drought analysis of past rainfall records of Tongatapu observed that the average duration of droughts was 37 months and they occurred on nearly every 11.5 years with a wide range in frequency and duration [22]. The longest period of zero recharge was 18 consecutive months for past rainfall records. It was also observed that the salinity of water pumped from Mataki'eua/Tongamai well field is most strongly correlated with the amount of rainfall received in the preceding 12 to 18 months [22,49]. Groundwater development from Mataki'eua/Tongamai well field began in 1966 (MLNR, Tonga 2017). The current groundwater withdrawal in the same well field is over eight times the extraction at the beginning of pumping (MLNR, Tonga 2017). Many wells in the well field use diesel pumps that pump at about 270 m$^3$/day. Possible use of electric pumps and the growing water demands can increase the pumping rates (MLNR, Tonga 2017) in the future.

Steady-state simulation results can indicate the potential impact and the worst case of the climate change scenario and have been used in many previous studies [13,50]. Current recharge value was progressively decreased to estimate the recharge rate at which well salinization started. While pumping rate was maintained at current value, well salinization occurred at about 69% of recharge. Similarly, pumping rate was increased from the current value to obtain pumping rate at which saltwater intrusion occurred in public wells. For current recharge values, 160% of present pumping initiated well salinization. Thus, in this study, six different scenarios (Table 1) are considered to simulate the possible impacts of changes in recharge and pumping on the freshwater lens and well salinization. Scenarios are selected to cover the increasing impact of drought and groundwater development conditions on well field salinization. The base scenario represents the calibrated model at base recharge and pumping conditions. The first three scenarios represent potential droughts, and the next three represent scenarios of increased groundwater development. Either recharge or pumping is varied from the base case in each scenario so that the relative impact of recharge and pumping can be compared.

**Table 1.** Long-term stresses scenarios.

| Sl.No | Scenario | Recharge (m/year) | Pumping from Public Wells (m³/day) |
|---|---|---|---|
| 1 | Base (current) | 0.570 (R) | 10,410 (P) |
| 2 | 0.69 R | 0.393 | 10,410 |
| 3 | 0.65 R | 0.371 | 10,410 |
| 4 | 0.60 R | 0.342 | 10,410 |
| 5 | 1.6 P | 0.570 | 16,656 |
| 6 | 1.8 P | 0.570 | 18,738 |
| 7 | 2.0 P | 0.570 | 20,820 |

## 3. Results

### 3.1. Results of Calibration

The hydraulic conductivity value of 3600 m/d was obtained as a result of manual calibration. The value is near the upper end of the values reported in the literature and close to that from pumping tests [22]. Statistical results of freshwater thickness calibration are given in Table 2. The simulated and observed freshwater thickness at salinity monitoring boreholes indicate a correlation coefficient of 0.87 with a normalized root mean square error of 13.8%.

**Table 2.** Statistical results of simulated freshwater thickness.

| Parameters | Values |
|---|---|
| Number of observation points | 16 |
| Correlation coefficient | 0.87 |
| Range in observations | 8.23 m |
| Maximum residual | 2.65 m |
| Minimum residual | −2.84 m |
| Residual mean | 0.27 m |
| Absolute residual mean | 1.15 m |
| Root mean squared error | 1.44 m |
| Normalized RMS error | 13.8% |
| Scaled RMS error | 17.5% |

Figure 3a shows the comparison of simulated freshwater thickness and observed freshwater thickness at the 16 SMBs. The contours based on average observed freshwater thickness at SMBs along with simulated freshwater thickness across the island are shown in Figure 3b. Spatial distribution of residuals at SMBs are also indicated in Figure 3b. The model captures the spatial distribution of freshwater thickness though there are differences at salinity monitoring boreholes. It is reported that the agricultural fields around SMB08 rely heavily on groundwater from village wells [22]. As all the village wells in this region have not been included in the model due to the lack of well data (MLNR Tonga, 2017), freshwater thickness may have been overestimated at SMB08, indicating a minimum residual of 2.84 m Fua'amotu region where SMB09 and SMB10 are located and receive higher rainfall than the rest of Tongatapu [22]. The non-uniform spatial distribution of recharge could have resulted in an underestimation of freshwater thickness by the model in the above region. The maximum residual of 2.65 m is at SMB10. The assumption of uniform recharge and pumping could have led to differences in freshwater thickness at other locations.

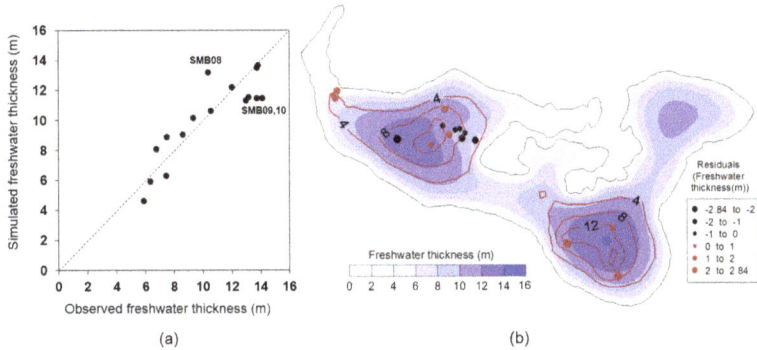

**Figure 3.** Steady-state calibration results; (**a**) simulated-versus-observed freshwater thickness at SMBs and (**b**) contours of simulated (blue) and observed (red) freshwater thickness; observed freshwater thickness contours are based on freshwater thickness shown in Figure 3a.

### 3.2. Current State of Freshwater Lens and Impact of Current Pumping

The current state of freshwater lens thickness is indicated in Figure 3b. The total volume of recharge is 147.6 MCM/year, and the total pumping rate is 6.1 MCM/year, which amounts to roughly 4.1% of the recharge. The maximum thickness is 14.1 m occurring around the Fua'amotu region, and the average thickness of freshwater is 8.4 m for the island. The regions with a freshwater lens thicker than 10 m are Fua'amotu, Liahona, and Kolonga region. Freshwater volume is calculated as the product of porosity and sum of the area of each element multiplied by freshwater thickness in the element. The estimated volume of freshwater contained in regions where the lens thickness is greater than 10 m is 294 MCM considering a porosity value of 0.3. The maximum freshwater head under sustained steady-state pumping is obtained as 0.329 m. At the current rate of pumping used in the study, it is observed that there is no seawater intrusion into both the public and village wells. Detailed statistics of the lens are specified in Table 3.

The impact of the current groundwater development on the predevelopment freshwater lens is investigated. Statistics of the freshwater lens for predevelopment and post-development conditions are presented in Table 3. Average freshwater head, maximum freshwater thickness, and minimum interface elevations decrease by about 5%. However, area and volume of the freshwater lens with thickness greater than 10 m decrease by 9% and 12%, respectively, for the post-development simulation. The impact of concentrated pumping from the public well field at Mataki'eua/Tongamai is evident in Figure 4 where the contours indicate a decrease in predevelopment freshwater thickness due to current pumping. Though the well field covers less than one km$^2$, the freshwater lens thickness reduced by over 0.5 m, 1 m, and 2 m in an area of 43 km$^2$, 19.3 km$^2$, and 12.3 km$^2$, respectively. Contours are asymmetric in shape and are concentrated along the coast as drawdown is restricted at the coast where specified head boundary conditions are defined. As the village wells are distributed across the island and have lower pumping rates effect of the village well pumping on the freshwater lens is not significant.

Table 3. Impacts of the current pumping condition.

| Parameters | | Current Pumping | Predevelopment | Percentage Decrease |
|---|---|---|---|---|
| Maximum freshwater head (m) | | 0.329 | 0.347 | 5.19 |
| Minimum interface elevation(m) | | −13.72 | −14.46 | 5.12 |
| Freshwater thickness (m) | Maximum | 14.05 | 14.81 | 5.13 |
| | Average | 8.40 | 9.02 | 6.87 |
| Area (km²) where lens thickness is | >5 m | 206.5 | 208.2 | 0.79 |
| | >7 m | 167.3 | 170.8 | 2.08 |
| | >10 m | 83.1 | 91.7 | 9.39 |
| Volume (MCM) of freshwater where lens thickness is | >0 m | 636.7 | 658.4 | 3.30 |
| | >5 m | 581.4 | 604.7 | 3.86 |
| | >7 m | 510.1 | 536.9 | 5.00 |
| | >10 m | 294.8 | 334.6 | 11.90 |

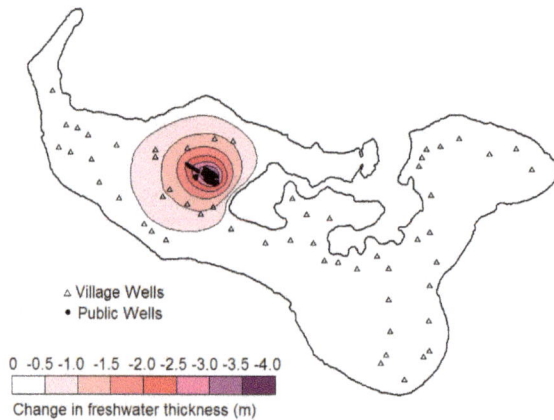

△ Village Wells
• Public Wells

0  -0.5 -1.0  -1.5  -2.0 -2.5 -3.0 -3.5 -4.0

Change in freshwater thickness (m)

**Figure 4.** Impact of current groundwater development on the predevelopment freshwater lens. Contours indicate the difference between post development and predevelopment freshwater thickness.

*3.3. Predicted Effects of Long-Term Stresses*

3.3.1. Impact on Freshwater Lens

The effect of long-term stresses is indicated as a change in the present state of the freshwater volume. The percentage of change in total freshwater volume (>0 m thick), the volume of freshwater where the thickness is more than 5 m, 7 m, and 10 m in comparison to the base case are indicated in Figure 5. Detailed freshwater statistics under scenarios of reduced recharge are presented in Table S2. The total volume of freshwater lens is decreased by 18% for scenario 0.69 R and by 24% for scenario 0.6 R. The deeper freshwater lens indicated by freshwater volume with lens thicker than 10 m decreased by 65% and 86% for scenarios 0.69 R and 0.6 R, respectively, indicating a higher sensitivity of the thicker part of the freshwater lens to applied stresses. The effect of increased pumping on freshwater lens volume change is much less compared to that of reduced recharge as pumping is concentrated over a smaller area. Detailed freshwater statistics under scenarios of increased pumping are presented in Table S3. The total freshwater volume decreased by 2% and 3% for scenario 1.6 P and 2 P, respectively. The part of freshwater lens thicker than 10 m resulted in reductions of 9% and 13% for scenarios 1.6 P and 2 P, respectively.

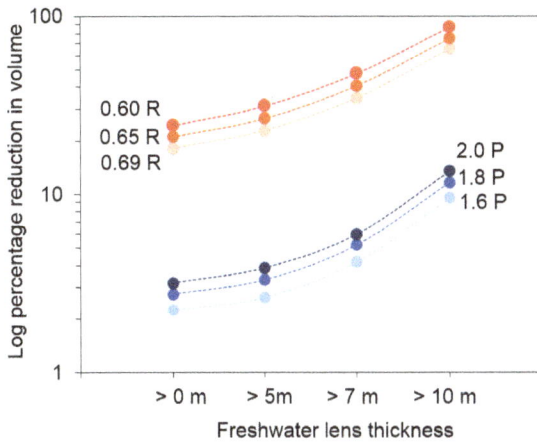

**Figure 5.** Reduction in volume of the freshwater lens compared to base case under various rates of recharge and pumping compared to base recharge of 570 mm/year (R) and current pumping of 16,710 m$^3$/day (P).

### 3.3.2. Well Salinization

Saltwater intrusion into wells is quantified in terms of the number of wells salinized and saltwater ratios in each well. With the best estimates of pumping rates and screen lengths, no well salinization was observed at the village well locations in the model, under the scenarios considered in this study.. Spatial distribution of saltwater intrusion into public wells under various scenarios of reduced recharge rates is shown in Figure 6. A saltwater ratio of 0.001 is considered to be the onset of well salinization. For scenario 0.69 R, two wells located at the center of the well field get salinized. With further reduction in recharge wells around the center and those near the lagoon are intruded by saltwater. Though freshwater thickness is reduced across the well field due to a decrease in recharge, concentrated pumping expedites saltwater intrusion into well field from the lagoon side, indicating lateral saltwater intrusions. Spatial distribution of saltwater intrusion into public wells under increased pumping is shown in Figure 7. As expected, wells located at the center of the well field get salinized first. Increased pumping rates cause saltwater upconing and more wells around the center are salinized.

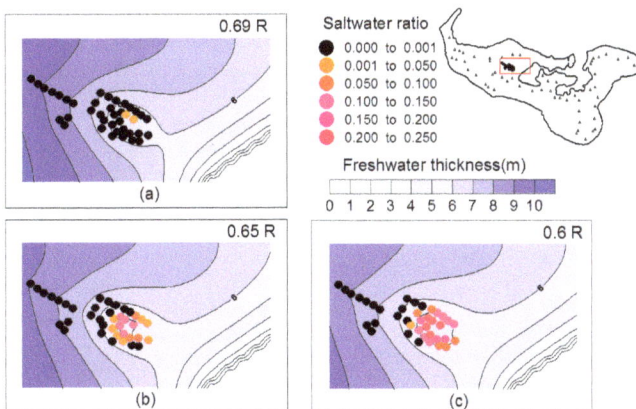

**Figure 6.** Saltwater intrusion into public wells under decrease in base recharge rates (R = 570 mm/year) (**a**) 69% of base recharge, (**b**) 65% of base recharge, and (**c**) 60% of base recharge.

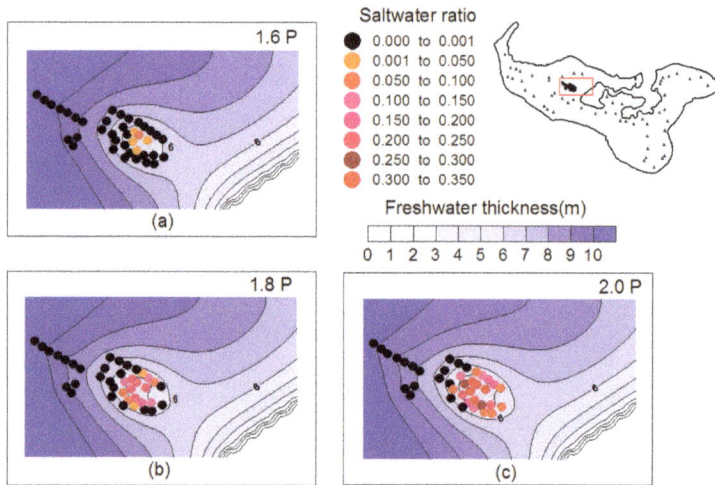

**Figure 7.** Saltwater intrusion into public wells under increase in current pumping (P = 16,710 m³/day) (a) 160% current pumping; (b) 180% current pumping; and (c) 200% current pumping.

The percentage of total wells intruded by saltwater, and maximum saltwater ratios for different scenarios are given in Table 4. For scenarios with reduced recharge 0.69 R, 0.65 R, and 0.6 R, saltwater intrusion occurs in two, 15, and 20 of the public wells, respectively. The maximum saltwater ratio was 0.21 for 0.6 R. While increased pumping causes saltwater intrusion in five, 15, and 21 public wells for 1.6 P, 1.8 P, and 2 P, respectively. Maximum saltwater ratios for the worst case pumping scenario (2 P) is 0.35, whereas it is 0.22 for the worst case recharge (0.6 R). Both the above scenarios cause around 50% of wells to be intruded by saltwater.

**Table 4.** Well salinization under different scenarios.

| Sl.No | Scenario | Percentage of Wells Intruded by Saltwater | Maximum Saltwater Ratio |
|---|---|---|---|
| 1 | Base (current) | 0.0 | 0.00 |
| 2 | 0.69 R | 5.1 | 0.004 |
| 3 | 0.65 R | 38.5 | 0.13 |
| 4 | 0.60 R | 51.3 | 0.22 |
| 5 | 1.6 P | 12.8 | 0.05 |
| 6 | 1.8 P | 38.5 | 0.24 |
| 7 | 2.0 P | 53.8 | 0.35 |

## 4. Discussion

The numerical model developed in this study revealed the quantitative behavior of freshwater lens and replicated observed freshwater thickness with a root mean square error of 1.44 m. Island aquifers are characterized by high heterogeneity and lack of data [7]. Previous modeling studies on real islands using dispersion models reported scaled-RMSE of 10% for Bonriki Island model with an area of 9.5 km² [12] and 21% for Kish Island model with an area of 112 km² [10] for modeled versus measured salinity. Scaled RMSE for modeled versus measured freshwater thickness is 17.5% in this study (Table 2). Given the large size of the island (259 km²) and simplifying assumptions, the model-measured difference that is not unacceptable.

The total volume of freshwater lens is estimated to be 637 MCM for an annual recharge of 147.6 MCM. Current groundwater development is estimated at 6.1 MCM/year. The volume of the freshwater lens with the thickness greater than 10 m is less than 300 MCM, and exist in less than one-third of the total area of the island (Table 3). It is seen that the current pumping rate of about 4.1% of annual recharge can be sustained by a freshwater lens with no saltwater intrusion under the assumption of uniform recharge throughout the year. However, its impact on predevelopment groundwater lens appears to be significant. The area and volume of the freshwater lens with thickness greater than 10 m are reduced by 9.4% and 11.9%, respectively, as compared with those for the predevelopment lens. Current pumping reduced freshwater thickness in an area of 12.3 km$^2$ around well field by over twometers, as compared to the predevelopment state of freshwater lens. Jakovovic et al. [51] adopted an interface rise of two meters to delineate a region around a pumping well as saltwater upconing zone of influence (SUZI). A limit of two meters is about 25% of the average freshwater thickness of Tongatapu. Considering the above criteria of the saltwater upconing zone of influence, an area over 20 times well field size indicates a very significant reduction in freshwater thickness in Tongatapu.

Scenarios considered in the study investigated the effect of drought and increased pumping. For a scenario with a 40% decrease in average annual recharge, 86% of freshwater lens volume with thickness more than 10 m is lost. For pumping at 8.2% of the average annual recharge, loss of freshwater lens volume with thickness more than 10 m is 13%. Groundwater development in Tongatapu is concentrated over a smaller area which leads to gradient reversals (ocean to land) resulting in saltwater intrusions (Figure 7). Salinization risk of each well is influenced by location from the center of the pumping well field and also the distance from the lagoon (Figures 6 and 7). For scenarios with a reduction in recharge (0.69 R, 0.65 R, and 0.6 R), lateral saltwater intrusions are prominent, whereas saltwater intrusion due to upconing under wells is significant for scenarios with increased pumping rates (1.6 P, 1.8 P, and 2 P). However, wells with the highest risk of salinization are almost common for scenarios of reduced recharge and increased pumping. Identification of high-risk wells helps in the planning and management of groundwater pumping.

Among the scenarios considered in this study, the worst cases of 0.6 R and 2 P cause over 50% of public wells to be intruded by saltwater (Table 4), even though total freshwater volume change is 24% for 0.6 R and 3% for 2 P, as compared to the base case (Figure 5). Hence, evaluation of freshwater volume change alone may not indicate the potential threats due to pumping. In scenario 2 P, though the pumping rate is about 8.2% of the average annual recharge, 50% of wells are intruded by saltwater. Well salinization of over 50% of number of public wells indicates a reduction in freshwater production by 50%. This further emphasizes the need for groundwater management and planning.

To evaluate the relative impact of pumping and recharge, the total volume of water extracted is quantified as the reduction in recharge [22]. This approach can also evaluate the relative impact of pumping distributed across total area and localized pumping. As base pumping rates are about 4% of recharge, scenario 1.8 P represents a 7% decrease in recharge. Well salinization, in terms of the number of wells intruded, under this scenario, is comparable to that of 0.65 R where effective recharge reduction is 39%, including pumping. Similarly, scenario 2 P and 0.6 R are quite comparable in terms of the number of wells salinized. It is seen that pumping impacts on well salinization are about five times that of recharge for Tongatapu Island. The high relative contribution of pumping to well salinization can be attributed to the high concentration of wells in the small area in Tongatapu.

Since seawater intrusion into pumping wells is a local phenomenon, sustainable groundwater development from a limited number of pumping wells is much smaller than the estimated volume of freshwater. White et al. [22] estimated island-wide sustainable groundwater development to be 20% of groundwater recharge and total sustainable groundwater yield considering areal pumping rates as 54,000 to 72,000 m$^3$/d. The numerical model, considering the current distribution of pumping wells, indicates that extractions at about 6.5% of the annual average recharge, and 50% of the lower range of areal pumping rates, as mentioned above, can cause saltwater intrusions.

Even though the precipitation exhibits seasonal and yearly patterns, the steady-state assumption was not too unrealistic as the groundwater system responded quickly due to large hydraulic conductivity. The salinity monitoring wells in Tongatapu use multi-nested tubes resulting in different values of heads in each tube due to the density differences in the tubes. The mean water table elevation for Tongatapu from observations in village wells for 1971–2007 was 0.41 m above mean sea level [22].

Error in deducing head from multi-nested monitoring wells might be large, as compared to the head above the MSL at the location. Hence, careful attention is required to make full use of the water level observations from multi-nested tubes.

Spatial heterogeneity of the aquifer and recharge distribution have not been considered. The tidal effects were not included in the study. The model developed considers the sharp interface at the freshwater limit. Hence, the fluid above the interface is freshwater, but in reality, the fluid below is not completely saltwater. Error of neglecting mixed saltwater below the interface on general flow pattern needs to be investigated further. While all the public wells for urban water supply have been included in the model, only 56 village wells were represented out of the over 200 village wells known to exist [36] as there was no information about the location of other wells. Estimated total groundwater development rate of the island was applied uniformly to all wells in the well field as individual pumping rates were not known. A more accurate representation of wells, in terms of locations, pumping rates, and screen length could result in better estimates of well salinization.

## 5. Conclusions

Prediction of the extent of freshwater lens and quantification of the local effects of pumping is essential for the design and development of effective water management strategies in small islands. In this study, a steady-state sharp-interface numerical model of fresh and saline groundwater flow is developed to evaluate saltwater intrusion and well salinization in Tongatapu Island. The model has a comparatively fine grid size of 25 m × 25 m in public well field and 100 m × 100 m in other areas, which is able to better represent the freshwater distribution and saltwater intrusion characteristics at pumping wells. Freshwater thickness is used as the calibration target in this study with the sharp interface corresponding to the upper limit of freshwater electrical conductivity adopted in the Kingdom of Tonga. Such an approach has not been reported in previous applications of sharp interface numerical models for islands. Thus, the current study is the first, to the best of our knowledge, to conduct a comprehensive analysis of freshwater lens and well salinization using sharp interface approach in real islands.

Model results indicate that at present conditions, 46% of the total freshwater volume of Tongatapu, estimated around 630 MCM, is from freshwater lens thicker than 10 m and concentrated in two regions towards the center of the island. This may indicate sufficient availability of resources for groundwater development. However, the concentration of pumping wells in a single well field increases the risk of saltwater intrusion. Various scenarios of decreased recharge and increased pumping were considered to evaluate saltwater intrusion and well salinization. More than 50% of public wells are intruded by saltwater for 40% decreased recharge (0.6 R) or groundwater pumping at 8% of average annual recharge (2 P). It is seen that pumping impacts on well salinization are about five times that of recharge for Tongatapu Island. The higher relative contribution of pumping to well salinization can be attributed to the high concentration of wells in the small area. The results suggest that the risk of salinization for individual well depends on the distance from the center of well field and the distance from lagoon or coast. However, the closeness of the well to the center of well field results in higher saltwater ratios and the extent of upconing. Major factor leading to well salinization was saltwater upconing rather than lateral saltwater intrusions. Hence, it is imperative to include pumping wells in a groundwater model, especially when pumping is localized.

While the model is currently limited to steady state analyses, it can help water managers of the island to assess potential impacts on groundwater resources caused by changes in some important hydrogeologic parameters, such as pumping rates and recharge rates. It also helps to identify wells with a high risk of salinization. Currently, an unsteady model is being developed to simulate drought conditions. An enhanced model will be linked with an optimization method to identify the best groundwater management schemes under drought conditions. The study highlights the applicability of a sharp interface numerical model calibrated for freshwater thickness to assess the extent of freshwater lens, saltwater intrusion, and well salinization in real islands. It can be used

*Water* **2018**, *10*, 1636

as the first stage analysis or when sufficient data and computational resources are not available for implementation of dispersion models.

**Supplementary Materials:** The following are available online at http://www.mdpi.com/2073-4441/10/11/1636/s1, Table S1: WATBAL parameters, Table S2: Freshwater resources for decreased recharge rates and percentage decrease compared to base scenario, Table S3: Freshwater resources for increased pumping and percentage decrease compared to the base scenario.

**Author Contributions:** Conceptualization, R.B. and N.P.; Formal analysis, R.B.; Funding acquisition, N.P.; Methodology, R.B. and N.P.; Project administration, N.P.; Resources, S.Y. and T.K.; Supervision, N.P.; Writing—original draft, R.B.; Writing—review & editing, N.P., S.Y. and T.K.

**Funding:** This research was supported by a grant (17AWMP-B066761-05) from AWMP Program funded by the Ministry of Land, Infrastructure, and Transport of Korean government.

**Acknowledgments:** Rainfall data for the Tongatapu Island was supplied by the Tonga Meteorological Services, salinity monitoring data and Digital Elevation Model (DEM) were supplied by Ministry of Lands and Natural Resources (MLNR), Kingdom of Tonga.

**Conflicts of Interest:** The authors declare no conflict of interest.

## References

1. Falkland, A.; Custodio, E.; Diaz Arenas, E.; Simler, E. *Hydrology and Water Resources of Small Islands: A Practical Guide*; UNESCO: Paris, France, 1991; pp. 24 & 208, ISBN 9231027530.
2. White, I.; Falkland, T. Management of freshwater lenses on small Pacific islands. *Hydrogeol. J.* **2010**, *18*, 227–246. [CrossRef]
3. Nurse, L.A.; Mclean, R.F.; Agard, J.; Briguglio, L.P.; Duvat-Magnan, V.; Pelesikoti, N.; Tompkins, E.; Webb, A. Small islands. In *Climate Change 2014: Impacts, Adaptation, and Vulnerability. Part B: Regional Aspects. Contribution of Working Group II to the Fifth Assessment Report of the Intergovernmental Panel on Climate Change*; Field, C.B., Barros, V.R., Dokken, D.J., Mastrandrea, M.D., Mach, K.J., Bilir, T.E., Chatterjee, M., Ebi, K.L., Estrada, Y.O., Genova, R.C., et al., Eds.; Cambridge University Press: Cambridge, UK, 2014; pp. 1613–1654, ISBN 9781107415386.
4. Cai, W.; Borlace, S.; Lengaigne, M.; Van Rensch, P.; Collins, M.; Vecchi, G.; Timmermann, A.; Santoso, A.; McPhaden, M.J.; Wu, L. Increasing frequency of extreme El Niño events due to greenhouse warming. *Nat. Clim. Chang.* **2014**, *4*, 111–116. [CrossRef]
5. Groundwater Monitoring in Small Island Developing States in the Pacific. International Groundwater Resource Assessment Centre (IGRAC), The Netherlands. Available online: http://ihp-hwrp.nl/wp-content/uploads/2016/11/Groundwater-Monitoring-in-Small-Island-Developing-States-in-the-Pacific.pdf (accessed on 30 September 2017).
6. Werner, A.D.; Alcoe, D.W.; Ordens, C.M.; Hutson, J.L.; Ward, J.D.; Simmons, C.T. Current practice and future challenges in coastal aquifer management: Flux-based and trigger-level approaches with application to an Australian case study. *Water Resour. Manag.* **2011**, *25*, 1831–1853. [CrossRef]
7. Werner, A.D.; Sharp, H.K.; Galvis, S.C.; Post, V.E.A.; Sinclair, P. Hydrogeology and management of freshwater lenses on atoll islands: Review of current knowledge and research needs. *J. Hydrol.* **2017**, *551*, 819–844. [CrossRef]
8. Bailey, R.T.; Jenson, J.W.; Olsen, A.E. Numerical modeling of atoll island hydrogeology. *Ground Water* **2009**, *47*, 184–196. [CrossRef] [PubMed]
9. Ketabchi, H.; Mahmoodzadeh, D.; Ataie-Ashtiani, B.; Simmons, C.T. Sea-level rise impacts on seawater intrusion in coastal aquifers: Review and integration. *J. Hydrol.* **2016**, *535*, 235–255. [CrossRef]
10. Ataie-Ashtiani, B.; Rajabi, M.M.; Ketabchi, H. Inverse modelling for freshwater lens in small islands: Kish Island, Persian Gulf: Freshwater lens in small islands. *Hydrol. Process.* **2013**, *27*, 2759–2773. [CrossRef]
11. Sulzbacher, H.; Wiederhold, H.; Siemon, B.; Grinat, M.; Igel, J.; Burschil, T.; Günther, T.; Hinsby, K. Numerical modelling of climate change impacts on freshwater lenses on the North Sea Island of Borkum using hydrological and geophysical methods. *Hydrol. Earth Syst. Sci.* **2012**, *16*, 3621–3643. [CrossRef]
12. Post, V.E.; Bosserelle, A.L.; Galvis, S.C.; Sinclair, P.J.; Werner, A.D. On the resilience of small-island freshwater lenses: Evidence of the long-term impacts of groundwater abstraction on Bonriki Island, Kiribati. *J. Hydrol.* **2018**, *564*, 133–148. [CrossRef]

13. Mahmoodzadeh, D.; Ketabchi, H.; Ataie-Ashtiani, B.; Simmons, C.T. Conceptualization of a fresh groundwater lens influenced by climate change: A modeling study of an arid-region island in the Persian Gulf, Iran. *J. Hydrol.* **2014**, *519*, 399–413. [CrossRef]

14. Ghassemi, F.; Alam, K.; Howard, K. Fresh-water lenses and practical limitations of their three-dimensional simulation. *Hydrogeol. J.* **2000**, *8*, 521–537. [CrossRef]

15. Sanford, W.E.; Pope, J.P. Current challenges using models to forecast seawater intrusion: Lessons from the Eastern Shore of Virginia, USA. *Hydrogeol. J.* **2010**, *18*, 73–93. [CrossRef]

16. Bailey, R.T.; Jenson, J.W.; Taborosi, D. Estimating the freshwater-lens thickness of atoll islands in the Federated States of Micronesia. *Hydrogeol. J.* **2013**, *21*, 441–457. [CrossRef]

17. Dausman, A.M.; Langevin, C.; Bakker, M.; Schaars, F. A Comparison between SWI and SEAWAT–the Importance of Dispersion, Inversion and Vertical Anisotropy. In Proceedings of the 21st Saltwater Intrusion Meeting (SWIM21), Azores, Portugal, 21–26 June 2010; de Melo, M.T.C., Lebbe, L., Cruz, J.V., Coutinho, R., Langevin, C., Buxo, A., Eds.; pp. 271–274.

18. Shi, L.; Cui, L.; Park, N.; Huyakorn, P.S. Applicability of a sharp-interface model for estimating steady-state salinity at pumping wells—Validation against sand tank experiments. *J. Contam. Hydrol.* **2011**, *124*, 35–42. [CrossRef] [PubMed]

19. Mehdizadeh, S.S.; Vafaie, F.; Abolghasemi, H. Assessment of sharp-interface approach for saltwater intrusion prediction in an unconfined coastal aquifer exposed to pumping. *Environ. Earth Sci.* **2015**, *73*, 8345–8355. [CrossRef]

20. Person, M.; Taylor, J.Z.; Dingman, S.L. Sharp interface models of salt water intrusion and wellhead delineation on Nantucket Island, Massachusetts. *Groundwater* **1998**, *36*, 731–742. [CrossRef]

21. Anderson, M.P.; Woessner, W.W.; Hunt, R.J. *Applied Groundwater Modeling: Simulation of Flow and Advective Transport*, 2nd ed.; Academic Press: Cambridge, MA, USA, 2015; pp. 380–385, ISBN 9780120581030.

22. White, I.; Falkland, A.; Fatai, T. *Vulnerability of Groundwater in Tongatapu, Kingdom of Tonga: Groundwater Evaluation and Monitoring Assessment, SOPAC/EU Reducing the Vulnerability of Pacific APC States*; Australian National University: Canberra, Australia, 2009.

23. Ghyben, B.W. Nota in verband met de voorgenomen putboring nabij, Amsterdam. *Hague* **1888**, *21*, 8–22.

24. Herzberg, A. Die wasserversorgung einiger Nordseebader. *J. Gasbeleucht. Wasserversorg.* **1901**, *44*, 824–844.

25. Huyakorn, P.S.; Wu, Y.S.; Park, N.S. Multiphase approach to the numerical solution of a sharp interface saltwater intrusion problem. *Water Resour. Res.* **1996**, *32*, 93–102. [CrossRef]

26. Bear, J.; Cheng, A.H.-D.; Sorek, S.; Ouazar, D.; Herrera, I. *Seawater Intrusion in Coastal Aquifers: Concepts, Methods and Practices*; Kluwer Academic Publisher: Dordrecht, The Netherlands, 1999; Volume 14, pp. 213–247, ISBN 0792355733.

27. Shi, L.; Cui, L.; Lee, C.J.; Hong, S.H.; Park, N.S. Applicability of a sharp-interface model in simulating saltwater contents of a pumping well in coastal areas. *J. Eng. Geol.* **2009**, *19*, 9–14.

28. White, I.; Falkland, A.; Fatai, T. Vulnerability of Groundwater Resources in Tongatapu. In Proceedings of the 34th World Congress of the International Association for Hydro-Environment Research and Engineering: 33rd Hydrology and Water Resources Symposium and 10th Conference on Hydraulics in Water Engineering, Brisbane, Australia, 26 June–1 July 2011; pp. 879–887.

29. Nath, D. *Tonga Water Supply System Description Nuku'alofa/Lomaiviti*; SPC Tech. Rep. No. 421; Secretariat of the Pacific Community: New Caledonia, France, 2010.

30. Hunt, B. An analysis of the groundwater resources of Tongatapu Island, Kingdom of Tonga. *J. Hydrol.* **1979**, *40*, 185–196. [CrossRef]

31. Bobba, A.G.; Singh, V.P. Prediction of Freshwater Depth due to Climate Change in Islands: Agati Island and Tongatapu Island. In Proceedings of the International Conference on Water, Environment, Ecology, Socio-Economics and Health Engineering (WEESHE), Seoul National University, Seoul, Korea, 18–21 October 1999; Singh, V.P., Seo, I.W., Sonu, J.H., Eds.; pp. 235–250.

32. The Kingdom of Tonga. *Tonga 2011 Census of Population and Housing, Volume 2: Analytical Report*; Tonga Department of Statistics, Tonga and the Secretariat of the Pacific Community: New Caledonia, France, 2011.

33. Furness, L.J.; Gingerich, S. Estimation of recharge to the fresh-water lens of Tongatapu, Kingdom of Tonga. *IAHS Publ.* **1993**, 317–322.

34. Hyland, K. *Expansion of the Salinity Monitoring Network across Tongatapu*; Ministry of Lands and Natural Resources: Nuku'alofa, Tonga, 2012.

35. Furness, L.J.; Helu, S.P. *The Hydrogeology and Water Supply of The Kingdom of Tonga*; Ministry of Lands and Natural Resources: Nuku'alofa, Tonga, 1993.

36. Van der Velde, M. Agricultural and Climatic Impacts on the Groundwater Resources of a Small Island: Measuring and Modeling Water and Solute Transport in Soil and Groundwater on Tongatapu. Ph.D. Thesis, UCL-Université Catholique de Louvain, Louvain, Belgium, May 2006.

37. Climate Risk Management for the Water Sector in Tonga. Available online: http://www.pireport.org/pireport/2015/July/Tonga_Paper_revised.pdf (accessed on 30 July 2016).

38. Pacific Climate Change Science Program Report. *Current and Future Climate of Tonga*; Australian Government: Canberra, Australia, 2014.

39. Waterloo, M.J.; Ijzermans, S. *Groundwater Availability in Relation to Water Demands in Tongatapu*; Dutch Risk Reduction DRR-Team Mission Report: Amsterdam, The Netherlands, 2017.

40. Presley, T.K. *Effects of the 1998 Drought on the Freshwater Lens in the Laura Area, Majuro Atoll, Republic of the Marshall Islands, Sci. Invest. Rep. 2005-5098*; US Geological Survey: Reston, VA, USA, 2005.

41. Bear, J. *Hydraulics of Groundwater*, 1st ed.; McGraw-Hill: New York, NY, USA, 1979; pp. 381–384, ISBN 978-0-486-45355-2.

42. Park, N.S.; Lee, Y.D. Seawater intrusion due to groundwater developments in eastern and central Cheju watersheds. *J. Korean Soc. Groundw. Environ.* **1997**, *4*, 5–13.

43. Hong, S.H.; Han, S.Y.; Park, N.S. Assessment of potential groundwater development in coastal area. *J. Korean Soc. Civ. Eng.* **2003**, *23*, 201–207.

44. Han, S.Y.; Hong, S.H.; Park, N.S. Assessment of coastal groundwater discharge for complex coastlines. *J. Korea Water Resour. Assoc.* **2004**, *37*, 939–947. [CrossRef]

45. Park, N.; Koh, B.R.; Lim, Y. Impacts of fresh and saline groundwater development in Sungsan watershed, Jeju Island. *J. Korea Water Resour. Assoc.* **2013**, *46*, 783–794. [CrossRef]

46. Hong, S.H.; Shi, L.; Cui, L.; Park, N. Artificial injection to control saltwater intrusion in groundwater-numerical study on a vertical cross section. *Eng. Geol. J.* **2009**, *19*, 131–138.

47. Jung, E.T.; Lee, S.J.; Lee, M.J.; Park, N. Effectiveness of double negative barriers for mitigation of seawater intrusion in coastal aquifer: Sharp-interface modeling investigation. *J. Korea Water Resour. Assoc.* **2014**, *47*, 1087–1094. [CrossRef]

48. PACC Technical Report. *Design for Improved Water Supply and Water Management in Hihifo District, Tonga, No. 7*; Secretariat of the Pacific Regional Environment Programme: Apia, Samoa, 2014.

49. van der Velde, M.; Javaux, M.; Vanclooster, M.; Clothier, B.E. El Niño-Southern Oscillation determines the salinity of the freshwater lens under a coral atoll in the Pacific Ocean. *Geophys. Res. Lett.* **2006**, *33*, L21403. [CrossRef]

50. Sefelnasr, A.; Sherif, M. Impacts of seawater rise on seawater intrusion in the Nile Delta aquifer, Egypt. *Groundwater* **2014**, *52*, 264–276. [CrossRef] [PubMed]

51. Jakovovic, D.; Werner, A.D.; de Louw, P.G.; Post, V.E.; Morgan, L.K. Saltwater upconing zone of influence. *Adv. Water Resour.* **2016**, *94*, 75–86. [CrossRef]

*water*

MDPI

*Article*

# Impacts of Sea Level Rise and Groundwater Extraction Scenarios on Fresh Groundwater Resources in the Nile Delta Governorates, Egypt

**Marmar Mabrouk** [1,2,*], **Andreja Jonoski** [2], **Gualbert H. P. Oude Essink** [3,4] and **Stefan Uhlenbrook** [5,6]

[1]   Ministry of Water Resources and Irrigation, Nile Water Sector, P.O. Box 11471, 11728 Cairo, Egypt
[2]   IHE Delft, Department of Integrated Water Systems and Governance, P.O. Box 3015, 2601 DA Delft,
      The Netherlands; a.jonoski@un-ihe.org
[3]   Deltares, Unit of Subsurface and Groundwater Systems, Daltonlaan 600, 3584 BK Utrecht, The Netherlands;
      gualbert.oudeessink@deltares.nl
[4]   Department of Physical Geography, Utrecht University, Princetonlaan 8a, 3584 CB Utrecht, The Netherlands
[5]   Water Resources Section, Delft University of Technology, P.O. Box 5048, 2600 GA Delft, The Netherlands;
      s.uhlenbrook@un-ihe.org
[6]   World Water Assessment Programme, UNESCO, Villa La Colombella, Colombella Alta, 06134 Perugia, Italy
*   Correspondence: marmarbadr80@yahoo.com; Tel.: +20-1129353635 or +31-619741794

Received: 21 August 2018; Accepted: 9 November 2018; Published: 20 November 2018

**Abstract:** As Egypt's population increases, the demand for fresh groundwater extraction will intensify. Consequently, the groundwater quality will deteriorate, including an increase in salinization. On the other hand, salinization caused by saltwater intrusion in the coastal Nile Delta Aquifer (NDA) is also threatening the groundwater resources. The aim of this article is to assess the situation in 2010 (since this is when most data is sufficiently available) regarding the available fresh groundwater resources and to evaluate future salinization in the NDA using a 3D variable-density groundwater flow model coupled with salt transport that was developed with SEAWAT. This is achieved by examining six future scenarios that combine two driving forces: increased extraction and sea level rise (SLR). Given the prognosis of the intergovernmental panel on climate change (IPCC), the scenarios are used to assess the impact of groundwater extraction versus SLR on the seawater intrusion in the Delta and evaluate their contributions to increased groundwater salinization. The results show that groundwater extraction has a greater impact on salinization of the NDA than SLR, while the two factors combined cause the largest reduction of available fresh groundwater resources. The significant findings of this research are the determination of the groundwater volumes of fresh water, brackish, light brackish and saline water in the NDA as a whole and in each governorate and the identification of the governorates that are most vulnerable to salinization. It is highly recommended that the results of this analysis are considered in future mitigation and/or adaptation plans.

**Keywords:** saltwater intrusion; sea level rise; Nile Delta aquifer; fresh groundwater volume; extraction; Nile Delta governorates

## 1. Introduction

The Nile Delta in Egypt, along with its fringes, covers an area of approximately 30,000 km$^2$ [1]. It is occupied by the most populated governorates in Egypt. About 60% of Egypt's population lives in the Nile Delta region. Agriculture activities are predominant in the region due to the nature of the soil and the presence of an irrigation system [2]. The Nile Delta aquifer (NDA) is a vast aquifer located between Cairo and the Mediterranean Sea [3]. The NDA was formed by Quaternary deposits with a wide variety of hydrogeological characteristics and spatially varying salinity levels [4]. These deposits represent different aggradation and degradation phases that were usually accompanied by sea level changes [5]. Recent years have brought scientific evidence of increased groundwater salinization in NDA, predominantly driven by increased groundwater extraction [4].

Salinity in groundwater is a major quality hazard that limits its usage and affects the productivity of agricultural areas that depend on irrigation from groundwater wells [6]. The nature and properties of salinity were reviewed with a view to its management in Reference [7]. In coastal aquifers, such as the NDA, the salinity in groundwater is influenced by human interventions through excessive groundwater extraction, while saltwater intrusion (SWI) induced by sea-level rise (SLR) is also anticipated [8].

In north Kuwait, a freshwater aquifer polluted by saline seawater was modelled using a 3D numerical model [9]. The researchers declared that solute transport modelling has become a significant tool for analysing the groundwater quality, as it can provide insight into past and present behaviour and predict water quality management scenarios. It was also highlighted that climate change is likely to lead to multiple stresses in the groundwater sector [10]. This research emphasized the need for the development of management models that simulate SWI in aquifers and the assessment of future compound groundwater challenges. Open challenges and uncertainties regarding the influences of climate change in coastal aquifers were identified in Reference [11]. In Kish Island, Iran, the combined impacts of SLR with the associated land inundation and climate-induced variation of natural recharge to the aquifer system were analysed [12]. The results showed that the combined impact of SLR-induced land inundation and recharge rate variation is more significant compared to SLR impacts alone. A review of SLR impacts on SWI in coastal aquifers together with other factors, such as recharge rate variation, land inundation due to SLR, aquifer bed slope variation and changing landward boundary conditions [13], concluded that the impacts of these combined factors on SWI need to be further investigated. The NDA, like other coastal aquifers, is subjected to salinization threats.

Regarding NDA, [14] focused on a stretch between Ras El Bar and Gamasa along Egypt's Northern coast. They found that the effect of SLR will be salinization of the NDA and they proposed artificial recharge through injection wells as a mitigation strategy. Further analysis of the impacts of SLR on SWI in the NDA was done by [15]. They concluded that large coastal areas will be totally submerged by SLR and that the shoreline will be shifted several kilometres inland. The research utilized a two-dimensional FEFLOW model [16]. This modelling study used many assumptions due to a lack of available data. A 3D modelling approach was used by [17] to identify the effect of the increasing SLR, decreasing surface water level and increasing groundwater extraction on SWI in NDA. They found that under a combined scenario of increased groundwater extraction rates by 100%, an increased SLR by 100 cm and a decreased surface water level by 100 cm would cause the SWI to extend 79.5 km in the west and 92.75 km in the east from the shoreline by 2100.

While these studies have addressed some aspects of the SWI problem in the NDA, research covering the whole Nile Delta using 3D modelling to predict future scenarios of SWI has been quite limited. This approach is used in the current article to analyse the combined impact of SLR and excessive groundwater extraction in the NDA. Unlike previous studies that mainly presented the landside shift of the shore boundary and the dispersion zone due to SLR, the advantage of using a 3D model for this analysis is that it allows the determination of the full spatial distribution of SWI and the analysis of volumes of available fresh groundwater under different future scenarios. This enables a comparative analysis of the available volumes of different groundwater types (fresh water, light brackish, brackish and saline water) under different scenarios for the whole NDA. Moreover, for the first time, the volumes of each groundwater types are analysed per Nile Delta governorate (administrative regions in Egypt), according to their location in the Nile Delta. This may lead to more location-driven recommendations for mitigation and adaptation measures that could be implemented at the governorate level.

To carry out this analysis, a 3D variable-density groundwater flow model coupled with salt transport developed using the SEAWAT modelling system [18]. Because this is a regional model that is focused on freshwater resources in the NDA, it does not consider the dynamic coastline or the local coastal flooding processes. Thus, it is assumed that the horizontal position of the hydraulic head boundary at the coast is fixed. The model was calibrated by the groundwater salinity conditions in 2010, the year that serves as a baseline model and reference condition for future scenarios because most of the data required for a reliable analysis is sufficiently available for that year. This analysis proposes six scenarios to assess the impact of SLR versus groundwater extraction in the Nile Delta and determines which factor is causing increased groundwater salinization. The values assumed for the SLR and groundwater extraction rates are based on IPCC reports [19] and the Egyptian National Plan, as will be discussed later, in detail. In the first scenario (Sc.1), the model run until the year 2500 with no increase in SLR or groundwater extraction (same values as in 2010) to test the influence of time only on the complex groundwater system in NDA (as SLR will not stop after 2100) and to investigate the natural autonomous salinization process [20]. The comparative analysis for the remaining five scenarios is carried out by running the model until the year 2100 with varying SLR and groundwater extraction rates. This analysis leads to a proposal of future groundwater extraction levels for different Nile Delta governorates, which can be considered in future plans for the overall development of groundwater resources in the area.

## 2. Study Area

The study area is the Nile Delta in the Northern part of Egypt (Figure 1). It is the most fertile region in Egypt and is surrounded by highly arid desert. The Nile Delta begins approximately 20 km north of the Cairo governorate in the south and extends to the Mediterranean Sea in the north, covering an area of about $30 \times 10^3$ km$^2$. In the west, it is bounded by the Alexandria governorate and in the east, by Port Said. The Nile Delta contains 11 governorates that have economic, cultural and agricultural importance to Egypt. The locations and the names of the Nile Delta governorates are presented in Figure 1.

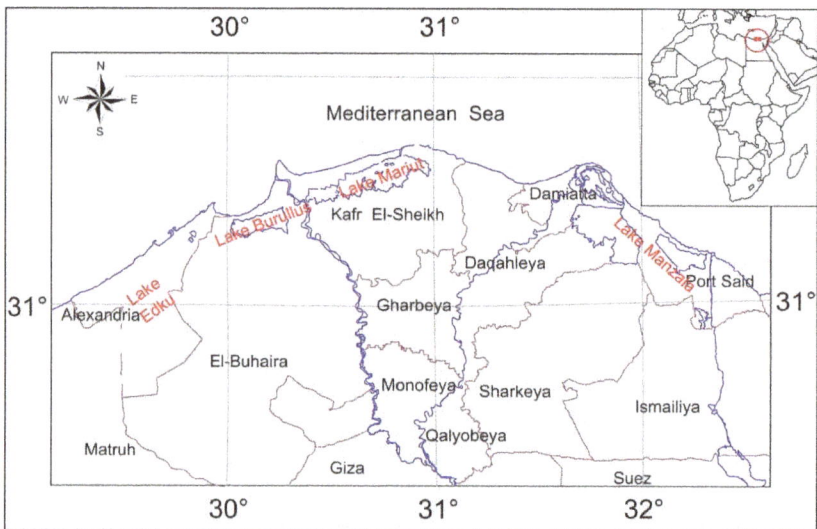

**Figure 1.** The study area of the Nile Delta.

The NDA is a large, semi-confined aquifer. It has Quaternary deposits that are classified into Holocene and Pleistocene strata [21]. The average thickness of the Holocene is 25 m. It reaches around 50 m close to the sea and vanishes towards delta fringes in the South. The Pleistocene is the main aquifer of the Nile Delta [22]. Its thickness varies from 200 in the south to 1000 m in the north. It is composed of sand and gravel with occasional clay lenses and it is underlain by a Pliocene layer composed of marine limestone and shale [2,3,21].

## 3. Method

### 3.1. Numerical Model

A 3D variable-density groundwater flow and coupled salt transport model for the NDA was developed using SEAWAT [18]. The model captures the situation in the year 2010, since for this year, most data is available. It includes a large amount of different types of hydrogeological data collected from several sectors and research institutions within the Egyptian Ministry of Water Resources and Irrigation (MWRI).

The model domain is discretized using 100 rows and 150 columns horizontally (grid cell sizes $2 \times 2$ km$^2$) and 21 modelling layers in the vertical direction, which enables sufficiently detailed simulations of salinity variations to be carried out under different conditions. The Mediterranean Sea is represented as a constant head and constant salinity boundary. The Suez Canal in the east is considered to be a no-flow boundary. The last (deepest) modelling layer of the 21 layers represents the impermeable Pliocene, which underlays the Pleistocene aquifer. This last modelling layer of the 21 layers also contains salinity sources arising from the dissolution of marine deposits present in the Pliocene [23]. The first (shallowest) modelling layer of the 21 layers represents the Holocene layer which is specified as having a constant horizontal hydraulic conductivity of 0.25 m/day. For the next 19 modelling layers, the horizontal hydraulic conductivity values vary from 15 to 150 m/day. The anisotropy of hydraulic conductivity is considered to be 10% [24]. The effective porosity for the Holocene is specified as having a constant value of 40% and it varies from 12% to 28% in the Pleistocene. The main source of recharge in the NDA is the excess irrigation water at the agricultural zones. This spatially varying recharge, together with spatially varying water levels and salinity concentrations in the main irrigation canals, was specified to the model to ensure proper characterization of the groundwater entering the modelling domain. The majority of canals have water with a maximum salinity of 0.3 kg/m$^3$ but in some locations, the salinity concentration reaches 0.65 kg/m$^3$ and this is even greater towards the Mediterranean. The overall groundwater extraction in 2010 was estimated to be about $4.9 \times 10^9$ m$^3$/year [4]. The wells are distributed according to their depth in the corresponding modelling layer.

The model was calibrated with observed salinity data coming from 155 observation wells. The calibration results provided a Root Mean Square Error (RMSE) of 0.2 kg/m$^3$. The absolute difference and the standard deviation between the simulated and the observed concentrations of the salinity in the Nile Delta were calculated to be 0.14 kg/m$^3$ and 0.11 kg/m$^3$, respectively. These values are quite small, so the developed model is considered to be sufficiently reliable for future analyses. Further details about the model development and calibration results are available in Reference [18].

### 3.2. Future Scenarios

One of the most important factors that control the development scenarios is the population growth rate, as it represents the main condition that directly affects future economic and social progress. According to the Egyptian Central Authority for Public Mobilization and Statistics [25], the population in Egypt has risen by approximately from 52 million capita in 1990 to 84 million capita in 2010. The growth rate decreased from 2.75% per year in the period 1976–1986 to 2.02% per year during the period 1996–2006. The total population is predicted to reach 200 million capita in 2100 [26]. These population growth rates have led to corresponding increases in water needs for different water-consuming sectors. Moreover, in the future, the ascending gap between water demands and supply will result in more tension in negotiations over transboundary water projects on the Nile [27]. Groundwater extraction rates are likely to increase over time too, assuming the continuation of the almost linear trend of increase in groundwater extraction that follows population growth, as presented in Figure 2. This trend of increased groundwater extraction was constructed based on data collected from the past, starting in 1980 and then projecting until 2100 [4].

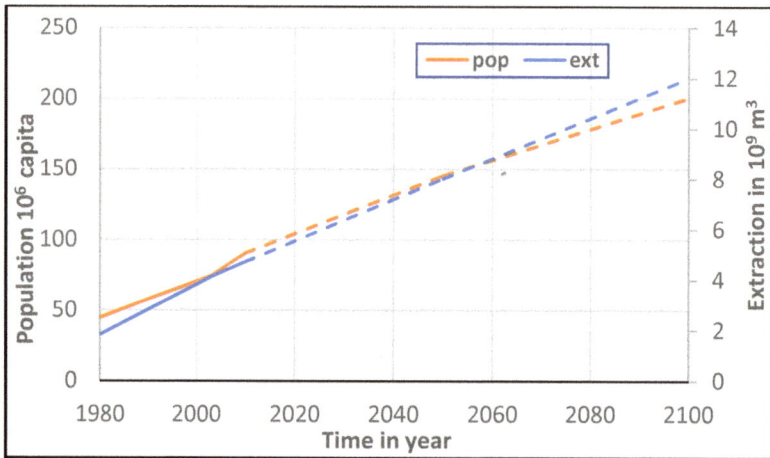

**Figure 2.** Population and groundwater extraction in the NDA as a function of time: current trend to 2010 (solid lines) and projections to 2100 (dashed lines) ([26]).

It is expected that climate change will substantially raise the sea level. The Intergovernmental Panel on Climate Change [19] emphasized the potential significance of SLR and predicted the global mean SLR by the year 2100 to be in the range 0.26–0.82 m compared to the earlier foreseen 0.18–0.59 m [28]. It was reported in Reference [29] that melting processes in Antarctica could contribute to a SLR of $1.05 \pm 0.3$ m by 2100 under RCP 8.5. A probability density function of global SLR in 2100 was constructed in Reference [30]. It was found that the probability of having a SLR of more than 1.8 m is less than five percent, with a median probability of 0.8 m. Their findings were based on process models combined with experts' opinions. They also stated that other lines of evidence are needed to justify any higher estimates of SLR for 2100. Recently, the probability density functions for extreme scenarios of global SLR in 2100 based on extreme mass loss from ice sheets using numerical simulations with a process-based model have been provided [31]. The median of their probability distribution was 0.73 m. All these high SLRs will lead to severe impacts on SWI in coastal aquifers.

Future salinization of the groundwater resources in the NDA is a complex process and its prediction has a high degree of uncertainty. Consequently, different scenarios are proposed in this article to identify a wide spectrum of possible future adaptation strategies. These scenarios cover extreme projections to estimate future changes in the NDA in terms of the salinity distribution over the next hundred years as a function of SLR and groundwater extraction rates. Our estimated $12 \times 10^9$ m$^3$/year of groundwater extraction in 2100 (Table 1) is consistent with the projections in the future scenarios of the MWRI national plan. Regarding the SLR, the selected ranges in the future scenarios (see again Table 1) were chosen by considering the SLR reported in Reference [19] and the median of the probability density functions for extreme scenarios of global SLR in 2100 reported in the literature review stated above.

Sc.1 (long run) is different from all of the other proposed scenarios. There is neither a SLR, nor an increase in groundwater extraction (the same values as in 2010 are used). This scenario analyses the impact of time only (until 2500), focusing on the autonomous salinization process, as it is known that the salinity distribution in coastal groundwater systems lags behind the current boundary conditions [20,32,33]. Thus, this scenario is not used for comparisons with the other scenarios. The remaining five future scenarios are combinations of different SLR and groundwater extraction rates.

**Table 1.** Future scenarios for sea-level rise (SLR) and groundwater extraction.

| Scenario (Sc.) | SLR (m) | Extraction ($10^9$ m$^3$/Year) | Time (Year) |
|---|---|---|---|
| Reference | 0 | 4.9 | 2010 |
| Sc.1 Long run | 0 | 4.9 | 2500 |
| Sc.2 Extreme | 1.5 | 12 | 2100 |
| Sc.3 Moderate | 1 | 8 | 2100 |
| Sc.4 Restrictive | 0 | 4.9 | 2100 |
| Sc.5 High ext. | 0 | 12 | 2100 |
| Sc.6 High SLR | 1.5 | 4.9 | 2100 |

Sc.2 (extreme) was selected to show the impact of the combined effect of a large SLR and extreme groundwater extraction on the NDA. The proposed population will reach 200 million per capita [26]. As the population growth continues with no birth control, a significant increase in groundwater extraction rate is expected $12 \times 10^9$ m$^3$/year to fulfil the high water demands (Figure 2). No control over groundwater extraction is assumed. In this scenario, the SLR is assumed to be 1.5 m, which is also considered extreme based on the analyses in the literature review presented earlier.

Sc.3 (moderate) takes birth control and a reduced population growth rate into account. It is associated with moderate control of groundwater extraction. The population is estimated to be about 146 million capita [26]. According to the national future plan of MWRI in 2100, the government will have moderate investments in reclamation projects. The groundwater extraction will be $8 \times 10^9$ m$^3$/year. The government will control the unplanned groundwater extraction and benefit from the solar desalination plants that are planned to be constructed. A moderate SLR of 1 m is also assumed.

Sc.4 (restrictive) is a very optimistic scenario as it assumes that no increase in groundwater extraction will occur, due to control by high financial penalties for any non-authorized groundwater extraction. In this scenario, the government depends on water resources other than groundwater. This scenario was selected to show the results of completely prohibiting groundwater extraction. Also, it is assumed that there will be no SLR. In fact, this scenario has same conditions as Sc.1 (long run) but it is analysed until 2100.

In Sc.5 (high ext.), the government encourages investments and land reclamation which leads to a dramatic increase in groundwater extraction and hence, increases groundwater salinization. In this scenario, the groundwater extraction rates are proposed to reach again $12 \times 10^9$ m$^3$/year by 2100. This scenario was selected to examine the impact of groundwater extraction alone, so the assumed SLR is 0 m.

On the contrary, the objective of Sc.6 (high SLR) is to measure the impact of SLR only. The groundwater extraction rate is same as now (2010) and a SLR of only 1.5 m is assumed. Scenarios 5 and 6 help to identify which is the main driving factor in the SWI process: SLR or groundwater extraction.

All six scenarios consider that the Egyptian policy towards cooperation with the Nile basin countries will continue to be the same and consequently, water levels in canals and water use for agriculture will remain constant. The increase of groundwater extraction in different scenarios is without any spatial variation compared to 2010. This study analysed how NDA will develop in the future with current spatial distribution of wells. Each existing well field will gradually increase the groundwater extraction until 2100. Further studies may investigate different spatial distributions of groundwater extraction. The initial conditions of the model were taken from the calibrated model's results for 2010. The model was run for 90 years for scenarios 2, 3, 4, 5 and 6 to predict the future SWI in 2100. For Sc.1, the model was run for 490 years to predict the SWI in the year 2500.

## 4. Results and Discussion

### 4.1. The Whole Nile Delta

The total volume of groundwater in NDA up to the hydrogeological base was estimated to be approximately $4050 \times 10^9$ m$^3$ covering an area of approximately $30 \times 10^3$ km$^2$. This agrees with Reference [15] who estimated the groundwater volume of NDA to be around $3600 \times 10^9$ m$^3$ but with a smaller modelled area ($24 \times 10^3$ km$^2$). According to [34], groundwater can be classified into different types with respect to the salinity level. Fresh water has a salinity of 0–1 kg/m$^3$, light brackish has 1–5 kg/m$^3$, brackish has 5–30 kg/m$^3$ and saline water has salinity greater than 30 kg/m$^3$. Figure 3 shows the different proportions of the groundwater types in 2010, together with their spatial distribution within the Delta. Fresh water accounts for 37%, light brackish 10%, brackish 21% and saline water 32%. The volume of total available fresh water is approximately $1290 \times 10^9$ m$^3$ and the volume of light brackish water is about $421 \times 10^9$ m$^3$. Table 2 provides estimates of the groundwater volume in $\times 10^9$ m$^3$ occupied by the four groundwater types for the current conditions in 2010 and for the six analysed scenarios.

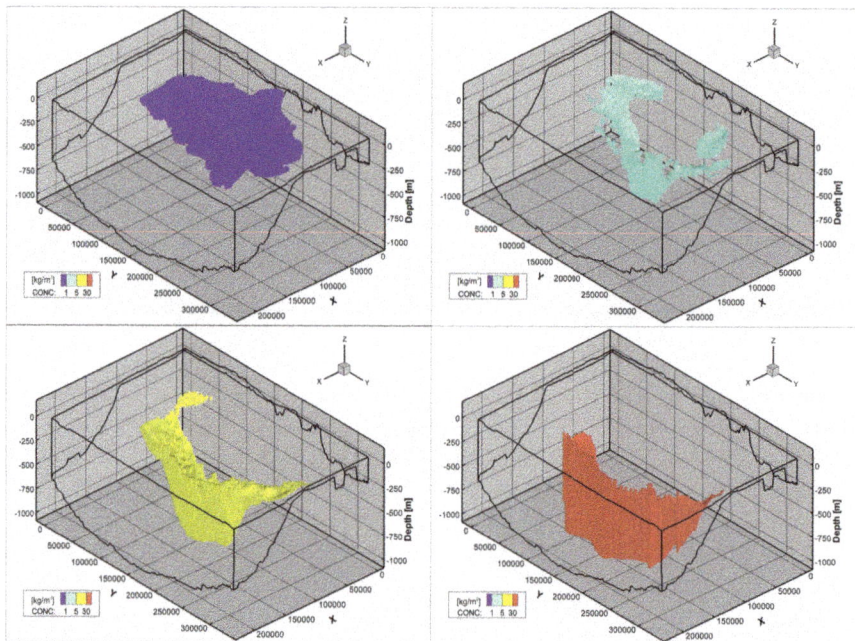

**Figure 3.** Spatial distribution of different groundwater types in the Nile Delta in 2010 (Concentration in kg/m$^3$).

It is clear from Table 2 that when only the influence of time is considered, the groundwater system becomes saltier. In Sc.1, the fresh groundwater volume decreases significantly by 31%, while the light brackish, brackish and saline groundwater volumes have corresponding increases. In this case, because of the long simulation period, the changes in salinization volumes are large in comparison with the other scenarios. The significance of the results from this scenario is that they demonstrate that even without any SLR or increased groundwater extraction, the salinization of the NDA will continue. It was shown by [18] that the NDA has not reached equilibrium yet and that this complex groundwater system is characterized by slow hydrogeological variations that bring significant impacts only after a long period of time [24]. For this reason, this scenario, in fact, raises a warning alarm. It shows the need for the planning of continuous adaptation strategies that prevent the accumulation of inland groundwater salinization. There are a number of adaptation strategies that could be applied in the Nile Delta [10], for example, artificial recharge, the extraction of saline and brackish groundwater and the modification of pumping practices through the reduction of withdrawal rates or adequate relocation of groundwater extraction wells.

**Table 2.** Groundwater volume in $\times 10^9$ m$^3$ in the Nile Delta aquifer (NDA) for the four groundwater types.

| Groundwater Types kg/m$^3$ | Current 2010 Reference | Sc.1 2500 Long Run | Sc.2 2100 Extreme | Sc.3 2100 Moderate | Sc.4 2100 Restrictive | Sc.5 2100 High Ext. | Sc.6 2100 High SLR |
|---|---|---|---|---|---|---|---|
| Fresh water 0–1 | 1290 | 893 −31% | 1049 −18.7% | 1119 −13.3% | 1190 −7.7% | 1090 −15.5% | 1147 −11.1% |
| Light brackish 1–5 | 421 | 436 +3.6% | 434 +3% | 432 +2.7% | 431 +2.4% | 433 +2.9% | 432 +2.8% |
| Brackish 5–30 | 829 | 1051 +26.8% | 900 +8.5% | 888 +7.1% | 886 +6.9% | 894 +7.9% | 890 +7.4% |
| Saline water >30 | 1513 | 1734 +14.6% | 1691 +11.7% | 1600 +5.7% | 1548 +2.2% | 1631 +7.7% | 1611 +6.4% |

For Scenarios 2 to 6, the models were run up until 2100 with different SLR rates and/or groundwater extraction rates. For Sc.2 (extreme), the fresh water decreased significantly by 18.7% in 90 years, while the saline volume increased by approximately 12%. These values are the highest among Scenarios 2 to 6, because Sc.2 shows the combined effect of an extreme SLR and groundwater extraction rate. Sc. 3 (moderate) shows modest values for fresh water volume loss (−13.3%) and saline groundwater gain (+5.7%).

Under very optimistic conditions, Sc.4 (Restrictive) has a small saline groundwater volume increase (+2.2%). The fresh groundwater loss (–7.7%) is the lowest among all other scenarios. To achieve this, however, very rigid control of groundwater extraction is necessary. Alternative, unconventional sources of water are required, possibly in combination with water saving and planting of crops that are more resistant to salt. Figure 4 shows the percent of change in different scenarios for the analysed groundwater types with respect to the 2010 baseline values.

In Sc.5 (high ext.), the fresh water volume decreases by about 15.5%. That decrease is second in severity after Sc.2 (extreme) with approximately 18.7%. This is also demonstrated in Figures 5 and 6 below, indicating that the largest part of the decrease in volume comes from groundwater extraction, not SLR.

Table 2 shows a decline in fresh water in Sc.6 (high SLR) to about 11%. This value is less than that in Sc.5 discussed above, meaning that SLR has a smaller effect on the groundwater volumes in the whole NDA compared to groundwater extraction.

Figure 5 shows different cross-sections of the NDA for Scenarios 2, 5 and 6. It is clear that the salinity front advances the most in Sc.2 (extreme) followed by Sc.5 (high ext.) and Sc.6 (SLR). The scenarios have the same ordering with respect to the size of the dispersion zone.

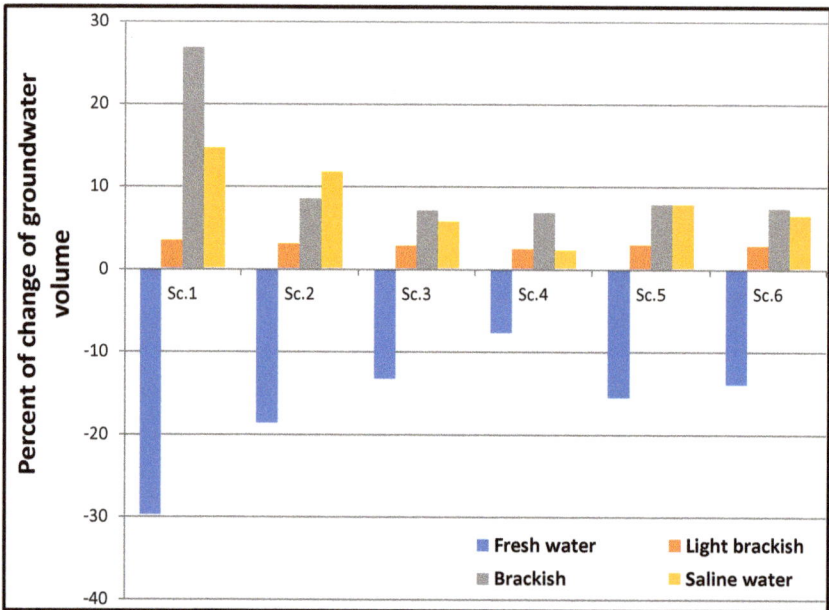

**Figure 4.** The percent of change of groundwater types in different scenarios, with respect to the 2010 baseline values.

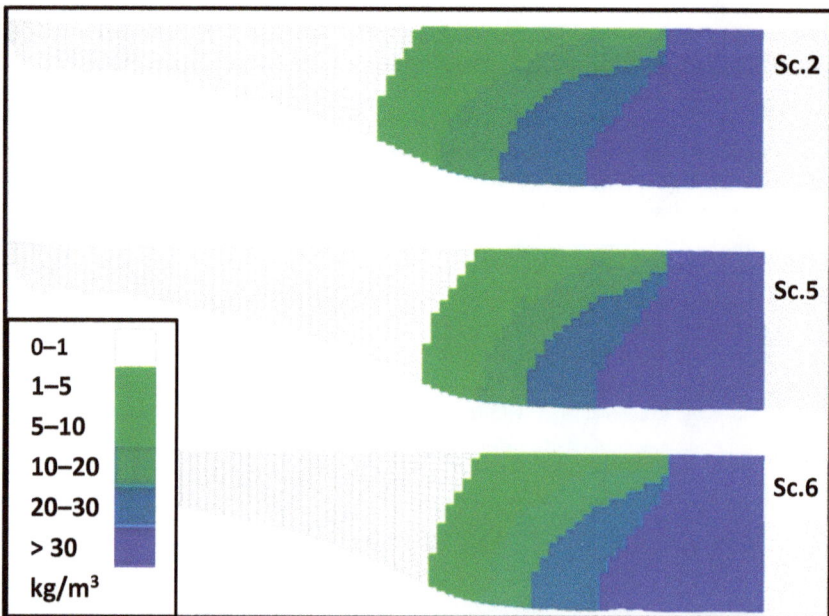

**Figure 5.** Cross-sections in the middle of the Nile Delta with different salinity distributions in different scenarios in the year 2100.

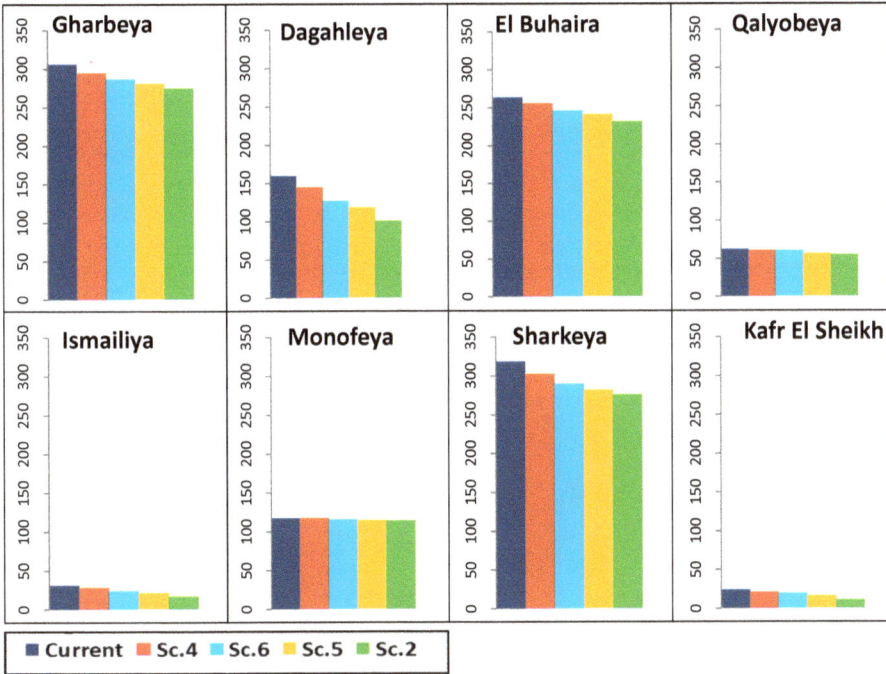

**Figure 6.** Block diagrams representing the fresh groundwater volume for different scenarios in selected governorates.

### 4.2. The Nile Delta Governorates

As stated in the study area section, the Nile Delta management system is organized into a number of governorates. The Egyptian governorates are administrative divisions. They are the second tier of the country's jurisdiction hierarchy, below the national government. Each governorate is administered by a governor, who is appointed by the President of Egypt [35]. The groundwater sector in Egypt is managed jointly by the MWRI and the governor in each governorate. Consequently, it is necessary to study salinization patterns with respect to each governorate, because each governorate has different hydrogeological parameters and more importantly, different agricultural and water resource management activities (groundwater extraction and irrigation) and thus, different groundwater type and salinization patterns.

Some of these governorates are coastal (e.g., Alexandria, Damietta), while others are far away from the zone where brackish and saline groundwater are present.

It is clear from Table 3 that El Buhaira, Gharbeya and Sharkeya have the highest freshwater volumes in all six scenarios (see Figure 1 for the locations of the governorates). On the other hand, coastal governorates have no fresh groundwater. Kafr El Sheikh and Ismailiya governorates have the lowest fresh groundwater volumes. The volume of fresh groundwater in each governorate is dependent on its location and distance from the coast, as well as the thickness of the aquifer, the governorate's area, the reclamation projects and the number of groundwater extraction wells. Here, again, it is clear that Sc.2 (extreme) has the highest impact of the governorates on salinization in 2100, followed by Sc.5 (high ext.) and Sc.6 (high SLR).

Figure 6 shows that for all scenarios, the fresh groundwater decreases at a lower rate in the Southern governorates, like Qalyobeya and Monofeya, compared to the Northern governorates, like Kafr El Sheikh. This is mainly because these governorates are far from the coastal zone. Meanwhile, the volume of fresh groundwater decreases at a high rate in the governorates Sharkeya and El Buhaira which suffer from the combined effect of excessive groundwater extraction and SLR. The analysis of the spatial distribution of salinity indicates that in these governorates, the Northern parts are already in a critical condition, while some continued groundwater extraction can still be allowed in the Southern regions. As we can see from Figure 6, Ismailiya governorate shows a significant drop of the (relatively small) fresh groundwater volume in all different scenarios. This governorate is already stressed by severe groundwater extractions, so future scenarios further intensify its criticality regarding available fresh water resources. In Dagahleya, the drop is also significant, although the absolute volume of fresh groundwater is larger. Coastal governorates (Kafr El Sheikh, Damietta and Alexandria) are very vulnerable to SLR. Damietta already has no fresh groundwater, while Kafr El Sheikh shows a severe reduction of fresh groundwater in the future in all scenarios.

**Table 3.** The fresh groundwater volume in $10^9$ m$^3$ with respect to different scenarios in all governorates of the Nile Delta.

| Governorate | Area $10^3$ km$^2$ | Extraction 2010 $10^6$ m$^3$/year | Current 2010 | Sc.1 Long Run | Sc.2 Extreme | Sc.3 Moderate | Sc.4 Restrictive | Sc.5 High ext. | Sc.6 High SLR |
|---|---|---|---|---|---|---|---|---|---|
| El-Buhaira | 10.130 | 1931 | 258 | 180 | 228 | 253 | 250 | 234 | 243 |
| Daghleya | 3.5 | 114 | 160 | 62 | 100 | 135 | 143 | 118 | 127 |
| Damietta | 1.029 | 362.5 | 0 | 0 | 0 | 0 | 0 | 0 | 0 |
| Gharbeya | 1.942 | 291.7 | 306 | 200 | 274 | 289 | 294 | 279 | 286 |
| Ismailiya | 2.10 | 163.8 | 31 | 8 | 16 | 26 | 28 | 20 | 24 |
| Kafr El Sheikh | 3.437 | 0.9 | 24 | 0 | 11 | 19 | 20 | 16 | 18 |
| Monofeya | 2.543 | 791.5 | 120 | 105 | 113 | 117 | 118 | 114 | 116 |
| Qalyobeya | 1.124 | 408.2 | 66 | 51 | 54 | 61 | 63 | 56 | 60 |
| Sharkeya | 4.18 | 681.5 | 318 | 210 | 276 | 295 | 302 | 281 | 290 |
| Alexandria | 2.679 | 2.4 | 22 | 5 | 7 | 14 | 15 | 9 | 13 |
| Portsaid | 1.351 | 154.2 | 57 | 96 | 45 | 50 | 51 | 46 | 49 |

Figure 7 shows the distribution of different groundwater types for selected governorates in the current situation (2010) and in 2100 under Sc.2 (Extreme). The Monofeya governorate has no saline or brackish groundwater, while the coastal governorates, for example, Kafr El Sheikh and Damietta, contain mainly brackish and saline groundwater.

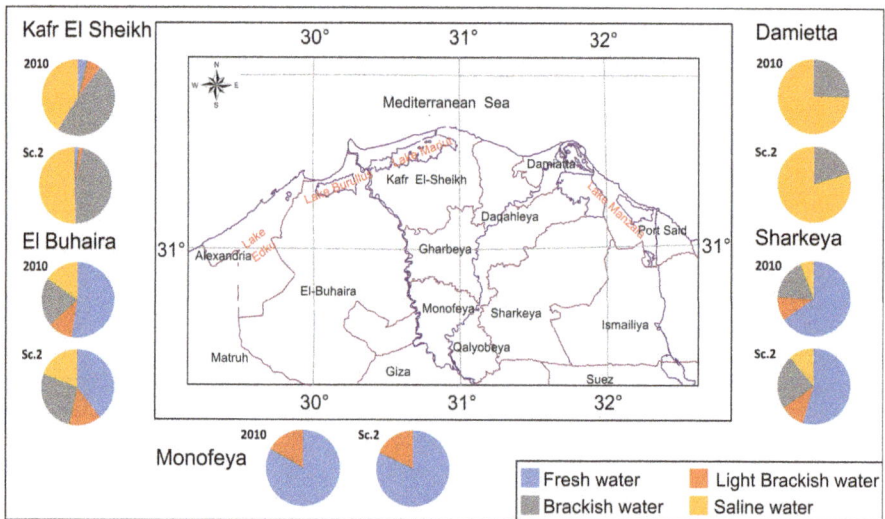

**Figure 7.** Proportional distribution of the different groundwater types for selected governorates in the current situation and under Sc.2

## 5. Conclusions and Recommendations

This article presented an assessment of the impacts of sea-level rise (SLR) and groundwater extraction on saltwater intrusion (SWI) in the Nile Delta aquifer (NDA). Six scenarios of different combinations of SLR and groundwater extraction from the NDA were proposed. The analysis was carried out using a 3D variable-density groundwater flow model coupled with salt transport. The groundwater salinity conditions were analysed by using the change in volume for the four groundwater types in the whole Nile Delta and for each Nile Delta governorate separately. The results clearly demonstrate that the effect of groundwater extraction is more severe than the SLR. However, SLR is linked to climate change, which is beyond the direct control of the Egyptian government and imposes an extra burden on the groundwater system of the NDA, especially when it is combined with excessive groundwater extraction. Salinization will intensify in the future if adaptation and mitigation measures are not implemented.

It is recommended that groundwater extraction is banned in Northern coastal governorates that are very sensitive to SLR and groundwater extraction. No groundwater extraction or investment in agriculture should be made in the coastal governorates Ismailiya and Dagahleya due to their shortage in fresh groundwater. In the Southern governorates, like Qalyobeya and Monofeya, the groundwater extraction of fresh water can continue, as the results indicate that these governorates have only fresh and light brackish water and do not have any brackish or saline water in all future scenarios. Groundwater extraction can be allowed only in the Southern parts of the Sharkeya and El Buhaira governorates, because their Northern parts are already in a critical condition. Table 4 summarizes the recommended locations for future groundwater extraction in different governorates of the Nile Delta.

**Table 4.** A summary of the recommended groundwater extraction locations in the Nile Delta governorates.

| Governorate | Recommendation | Reason |
|---|---|---|
| -Coastal governorates Kafr El Sheikh Alexandria Damietta Port Said | Ban groundwater extraction Search for alternatives | -Limited or no fresh groundwater -Sensitive to SWI (saltwater intrusion) |
| -Middle governorates El Buhaira Gharbeya Sharkeya | Extraction in the Southern region | -Fresh water is decreasing rapidly -Suffer from combined effects of excessive groundwater extraction and SLR -Have the highest fresh water volumes |
| Ismailiya Dagahleya | No groundwater extraction No investment in agriculture that relies on groundwater | -A huge drop of fresh water volume between different scenarios -SWI and ext. that intensify the criticality of its small fresh water volume |
| -Southern governorates Qalyobeya Monofeya | Groundwater extraction is recommended | -Fresh water is decreasing slowly -Far from SWI |

For strategic planning of groundwater management in Egypt until 2100, it is recommended that future groundwater extraction distribution is adjusted. The vulnerability of governorates in terms of the available fresh groundwater volume (under stress) should be taken into consideration when designing future water management adaptation plans. Taking serious measures, such as closing illegal wells, implementing full control of new authorization permits for groundwater extraction in the NDA and giving strict fines for any violation of groundwater extraction regulations, will be required to reduce the salinization problem. Additionally, future innovations and technologies, such as waste water reuse, crop management and desalinization, should be actively pursued. Moreover, immediate initialization of development projects that could protect groundwater salinization is needed, for instance, desalination and rain harvesting projects in the Northern coastal region.

Finally, it should be recognized that salt water intrusion is a long-term process which may led to the future deterioration of fresh groundwater resources in the NDA, even beyond 2100. Sc.1 which runs until 2500 indicates that time alone can bring further saltwater intrusion into the NDA. If uncontrolled development of groundwater resources continues while control and adaptation measures are not taken into account, this valuable fresh groundwater resource will be impaired to an extent that negatively affects the overall socioeconomic development of the country.

**Author Contributions:** The study is part of a PhD research conducted by M.M., under supervision of A.J., G.O.E. and S.U. The paper is conducted as a joint effort of all authors. Conceptualization, M.M., A.J. and G.O.E.; Methodology, M.M., A.J. and G.O.E.; Software, M.M.; Validation, M.M.; Formal Analysis, M.M., A.J. and G.O.E.; Resources, M.M.; Data Curation, M.M.; Writing-Original Draft Preparation, M.M.; Writing-Review & Editing, M.M., A.J. and G.O.E.; Supervision, A.J., G.O.E. and S.U.

**Funding:** This work has been supported financially by IHE Delft Institute for water education, Delft, The Netherlands.

**Acknowledgments:** The authors would like to thank the Ministry of Water Resources and Irrigation in Egypt for providing us with data.

**Conflicts of Interest:** The authors declare no conflict of interest.

## References

1. EGSA. *Egyptian General Survey and Mining: Topographical Map cover Nile Delta, Scale 1:2 000 000*; Egyptian General Survey and Mining (Publishing Center): Cairo, Egypt, 1997.
2. *SADS2030—Sustainable Agricultural Development Strategy*, 1st ed.; Ministry of Agriculture and Land Reclamation of Egypt: Cairo, Egypt, 2009; p. 197.
3. Farid, M.S.M. Nile Delta Groundwater Study. Master's Thesis, Cairo University, Cairo, Egypt, 1980.
4. Mabrouk, M.B.; Jonoski, A.; Solomatine, D.; Uhlenbrook, S. A Review of Seawater Intrusion in the Nile Delta Groundwater System—The basis for Assessing Impacts due to Climate Changes, Sea Level Rise and Water Resources Development. *Nile Water Sci. Eng. J.* **2017**, *10*, 46–61.
5. Diab, M.S.; Dahab, K.; El Fakharany, M. Impacts of the paleohydrological conditions on the groundwater quality in the northern part of Nile Delta, The geological society of Egypt. *Geol. J. B* **1997**, *4112*, 779–795.
6. Custodio, E. Aquifer overexploitation: What does it mean? *Hydrogeol. J.* **2002**, *10*, 254–277. [CrossRef]
7. Yihdego, Y.; Panda, S. Studies on Nature and Properties of Salinity across Globe with a View to its Management-A Review. *Glob. J. Hum. Soc. Sci. Res.* **2017**, *17*, 31–37.
8. Oude Essink, G.H.P.; Van Baaren, E.S.; De Louw, P.G.B. Effects of climate change on coastal groundwater systems, a modeling study in the Netherlands. *Water Resour. Res. J.* **2010**, *46*, W00F04. [CrossRef]
9. Yihdego, Y.; Al-Weshah, R.A. Assessment and prediction of saline sea water transport in groundwater using 3-D numerical modelling. *Environ. Processes J.* **2016**, *4*, 49–73. [CrossRef]
10. Gorelick, S.M.; Zheng, C. Global change and the groundwater management challenge. *Water Resour. Res. J.* **2015**, *51*, 3031–3051. [CrossRef]
11. Ojha, L.; Wilhelm, M.B.; Murchie, S.L.; McEwen, A.S.; Wray, J.J.; Hanley, J.; Massé, M.; Chojnacki, M. Spectral evidence for hydrated salts in recurring slope linear on Mars. *Nat. Geosci. J.* **2015**, *8*, 829. [CrossRef]
12. Mahmoodzadeh, D.; Ketabchi, H.; Ataie-Ashtiani, B.; Simmons, C.T. Conceptualization of a fresh groundwater lens influenced by climate change, A modelling study of an arid-region island in the Persian Gulf, Iran. *Hydrol. J.* **2014**, *519*, 399–413. [CrossRef]
13. Ketabchi, H.; Mahmoodzadeh, D.; Ataie-Ashtiani, B.; Simmons, C.T. Sea-level rise impacts on seawater intrusion in coastal aquifers: Review and integration. *Hydrol. J.* **2016**. [CrossRef]
14. Nofall, E.R.; Fekry, A.F.; El-Didy, S.M. Adaptation to the Impact of Sea Level Rise in the Nile Delta Coastal zone, Egypt. *Am. Sci. J.* **2014**, *10*, 17–29.
15. Sefelnasr, A.; Sheriff, M.M. Impacts of Seawater Rise on Seawater Intrusion in the Nile Delta Aquifer, Egypt. *Groundw. J.* **2014**, *52*, 264–276. [CrossRef] [PubMed]
16. Sherif, M.M.; Sefelnasr, A.; Javadi, A. Incorporating the concept of Equivalent Fresh water Head in Successive Horizontal Simulations of Seawater Intrusion in the Nile Delta aquifer, Egypt. *Hydrol. J.* **2012**, *464–465*, 186–198. [CrossRef]
17. Abdelaty, I.M.; Abd-Elhamid, H.F.; Fahmy, M.F.; Abdelaal, G.M. Investigation of some potential parameters and its impacts on saltwater intrusion in Nile Delta aquifer. *J. Eng. Sci.* **2014**, *42*, 931–955. [CrossRef]

18. Mabrouk, M.B.; Jonoski, A.; Oude Essink, G.H.P.; Uhlenbrook, S. Delineating the fresh-saline groundwater distribution in the Nile Delta using a 3D variable-density groundwater flow model coupled with salt transport. *Hydrol. J. Reg. Stud.* **2018**. under review.

19. Field, C.B.; Barros, V.R.; Dokken, D.J.; Mach, K.J.; Mastrandrea, M.D.; Bilir, T.E.; Chatterjee, M.; Ebi, K.L.; Estrada, Y.O.; Genova, R.C.; et al. *Climate Change 2014: Impacts, Adaptation, and Vulnerability. Part A: Global and Sectoral Aspects*; Contribution of Working Group II to the Fifth Assessment Report of the Intergovernmental Panel on Climate Change; IPCC: Geneva, Switzerland, 2014.

20. Delsman, J.R.; Hu-a-ng, K.R.M.; Vos, P.C.C.; De Louw, P.G.B.; Oude Essink, G.H.P.; Stuyfzand, P.J.; Bierkens, M.F.P. Paleo-modeling of coastal saltwater intrusion during the Holocene, An application to the Netherlands. *Hydrol. Earth Syst. Sci. J.* **2014**, *18*, 3891–3905. [CrossRef]

21. Sestini, G. Nile Delta: A review of depositional environments and geological history. *Geol. Soc. Lond. Spec. Publ.* **1989**, *41*, 99–127. [CrossRef]

22. Morsy, W.S. Environmental Management to Groundwater Resources for Nile Delta Region. Ph.D. Thesis, Faculty. of Engineering, Cairo University, Cairo, Egypt, 2009.

23. Van Engelen, J.; Oude Essink, G.H.P.; Kooi, H.; Bierkens, M.F.P. On the origins of hypersaline groundwater in the Nile Delta aquifer. *Hydrol. J.* **2018**, *560*, 301–317. [CrossRef]

24. Bear, J. *Hydraulics of Groundwater*; McGraw-Hill Book Company: New York, NY, USA, 1979; p. 592, ISBN 0-486-45355.

25. CAPMAS. *The Central Authority for Public Mobilization and Statistics, Egypt, Egypt in Numbers*; Ministry of Communication and Information Technology (Publishing Center): Cairo, Egypt, 2010.

26. United Nations, Department of Economic and Social Affairs, Population Division. World Population Prospects: The 2015 Revision, Key Findings and Advance Tables, United Nations, New York 2015, Working Paper No. ESA/P/WP.241. Available online: https://esa.un.org/unpd/wpp/publications/files/key_findings_wpp_2015.pdf (accessed on 15 November 2018).

27. Yihdego, Y.; Khalil, A.; Salem, H.S. Nile Rivers Basin Dispute: Perspectives of the Grand Ethiopian Renaissance Dam (GERD). *Hum. Soc. Sci. Res. J.* **2017**, *17*, 1–21.

28. Solomon, S.; Qin, D.; Manning, M.; Chen, Z.; Marquis, M.; Averyt, K.B.; Tignor, M.; Miller, H.L. *Contribution of Working Group I to the Fourth Assessment Report of the Intergovernmental Panel on Climate Change*; Cambridge University Press: Cambridge, UK; New York, NY, USA, 2007.

29. DeConto, R.M.; Pollard, D. Contribution of Antarctica to past and future sea-level rise. *Nature* **2016**, *531*, 591–597. [CrossRef] [PubMed]

30. Jevrejeva, S.; Grinsted, A.; Moore, J.C. Upper limit for sea level projections by 2100. *Environ. Res. Lett.* **2014**, *9*, 104008. [CrossRef]

31. Le Bars, D.; Drijfhout, S.; de Vries, H. A high-end sea level rise probabilistic projection including rapid Antarctic ice sheet mass loss. *Environ. Res. Lett.* **2017**, *12*, 044013. [CrossRef]

32. Meisler, H.; Leahy, P.P.; Knobel, L.L. *Effect of Eustatic Sea-Level Changes on Saltwater-Fresh Water in the Northern Atlantic Coastal Plain*; USGS Water Supply Paper; US Government Printing Office: Alexandra, VA, USA, 1984; Volume 2255, p. 34.

33. Larsen, F.; Tran, L.T.L.V.; Van Hoang, H.; Tran, L.T.L.V.; Christiansen, A.V.; Pham, N.Q. Groundwater salinity influenced by Holocene seawater trapped in incised valleys in the Red River delta plain. *Nat. Geosci.* **2017**, *10*, 376–381. [CrossRef]

34. Lide, D.R. (Ed.) *CRC Handbook of Chemistry and Physics*, 86th ed.; CRC Press: Boca Raton, FL, USA, 2015; p. 5585, ISBN 0-8493-0486-5.

35. Metz, H.C. (Ed.) *Egypt: A Country Study*; GPO for the Library of Congress: Washington, DC, USA, 1990.

*water*

MDPI

*Article*

# Saltwater Intrusion and Freshwater Storage in Sand Sediments along the Coastline: Hydrogeological Investigations and Groundwater Modeling of Nauru Island

**Luca Alberti, Ivana La Licata \* and Martino Cantone**

Dipartimento di Ingegneria Civile Ambientale, Politecnico di Milano, Piazza L. da Vinci 32, 20133 Milano, Italy; luca.alberti@polimi.it (L.A.); martino.cantone@polimi.it (M.C.)
\* Correspondence: ivana.lalicata@polimi.it; Tel.: +39-02-2399-6663

Received: 6 August 2017; Accepted: 9 October 2017; Published: 13 October 2017

**Abstract:** Water resources sustainable management is a vital issue for small islands where groundwater is often the only available water resource. Nauru is an isolated and uplifted limestone atoll island located in the Pacific Ocean. Politecnico di Milano performed a feasibility study for the development of sustainable use of groundwater on the island. This paper focuses on the first phase of the study that concerns the conceptual site model development, the hydrogeological characterization and the 2D model implementation. During the project, different activities were performed such as GNSS topographic survey of monitoring wells and groundwater level surveys taking into account tidal fluctuation. This data collection and the analysis of previous studies made it possible to identify the most suitable areas for groundwater sustainable extraction. The characterization findings suggested, unlike previous studies and surveys, the presence of only few drought resilient thin freshwater lenses, taking place in low conductivity sandy deposits, unexpectedly next to the seashore. Thanks to the 2D modeling results, it has been possible to clarify the mechanism that allows the storage of freshwater so close to the sea.

**Keywords:** atoll island; groundwater storage; freshwater resilience; MODFLOW/SEAWAT; saltwater intrusion; water resources management

---

## 1. Introduction

The integrated and sustainable management of water resources in small island countries is a fundamental issue for health and social well-being, protection of the environment and development of the economy. Here, it becomes a very high-level priority because of the strong limited nature of resources, worsened by climate change. Recently, White and Falkland [1] provided an insight into the key climatic, hydrogeological, physiographic, and management factors that influence the quantity of fresh water and saline intrusion into small islands aquifers. The Small Island Developing States (SIDS) organization has evaluated the sustainability problems faced by its member countries in the Pacific Regional Action Plan [2]. Falkland [3] and White et al. [4] described the main concerns that currently constrain the achievement of the goal of sustainable integrated water resources management in these countries: in many islands, there are insufficient hydrological data available for the analysis and planning of water governance reforms. Werner et al. [5] have shown in their review of the state of knowledge of atoll island groundwater that over fifty years of investigation have led to important advancements in the understanding of atoll hydrogeology, but a paucity of hydrogeological data persists on all but a small number of atoll islands. The lack of data is usually due to the difficulties related to the accessibility and the lack of instruments for parameter survey. Other issues include conflicts related to the use of water resources and location of water supply systems on customary land, problems with design and implementation of projects, and insufficient community education, awareness and participation.

Water resources on small islands can be classified as either conventional or non-conventional [6]. Conventional water resources include naturally occurring water such as surface water, groundwater, and rainwater. Non-conventional water resources involve a high level of technology and often high-energy consumption, such as desalination of seawater or brackish ground water, importation, or treated wastewater. Often, in small islands, lack of surface water and large periods of absence of precipitation make groundwater the only conventional and cheaper water resource available. Climatic projections suggest a dramatic amplification of droughts [7], with significant effects on groundwater availability, as observed in many fractured-karsts aquifers all over the world [8–10].

Groundwater occurs on small islands as either perched or basal aquifers [11]. The first ones occur over horizontal confining layers (aquicludes), while the second ones, which are the most common form on small coral and limestone islands, consist of unconfined, partially confined or confined freshwater bodies, which form at or below sea level. On many small coral and limestone islands, the basal aquifer takes the form of a "fresh water lens" which can underlie the whole island (Figure 1). These lenses of fresh groundwater accumulate from rainfall percolating through the soil zone and reside in fragile hydrodynamic equilibrium with the underlying saltwater, separated only by slight differences in density (transition zone). Basal aquifers are, indeed, vulnerable to saline intrusion owing to the freshwater-seawater interaction, and must be carefully managed to avoid over-exploitation and resultant increase of salt concentration in groundwater [6]. However, as shown in this study, freshwater in small islands can also store next to the seashore, in the sandy parts of the coast. The hydrogeology of atoll aquifers ("dual aquifer" systems in the hydrologic literature) is unique, consisting of a surficial, relatively low-permeability Holocene carbonate sand aquifer lying on a higher permeability Pleistocene limestone or karst aquifer [12]. The hydraulic conductivity of the Holocene aquifer, where the freshwater lens resides, is generally one to two orders of magnitude less than that of the underlying Pleistocene aquifer [13]. As reported by Nakada et al. [14], low permeability materials can suppress salinization and keep the groundwater relatively fresh and recharged by rainfall, even in low-elevation areas surrounded by saline water. Asymmetry is common in the lenses of atoll islands: the freshwater lens is commonly thicker on the lagoon side than on the reef side of the island and a cross-island variation in hydraulic conductivity is the usual explanation [15].

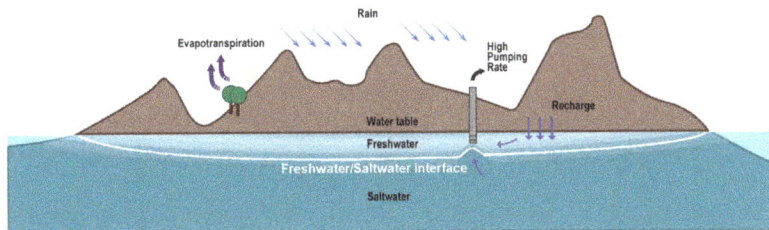

**Figure 1.** Small islands' typical freshwater lens-shape over the brackish/salt water. The effect of a pumping well (upconing) is indicated.

Through field investigations, long term surveys and groundwater modeling on Nauru island, this paper means to understand the hydrogeological structures and the groundwater process that allow fresh water to store in that position. This aspect needs to be clarified in order to quantify the fresh groundwater resource and to understand if these freshwater lenses are resilient to saltwater intrusion even in drought periods. Generally, resilience describes the capability of a system to maintain its basic functions and structures in a time of shocks and perturbations and can continue to deliver resources and ecosystem services that are essential for human livelihoods [16]. From the hydrogeological point of view, groundwater resilience defines the capability of the freshwater body to maintain its function of water supply even in periods of crisis as drought or not sufficient water supply from other resources. The time frame then should be the maximum crisis period length the hydrogeological system has so far experienced. The study of Nauru's hydrogeology is a step toward the knowledge of groundwater circulation into highly permeable aquifers, underlain and surrounded by seawater, and could contribute to improving the global knowledge on water management in these particular and fragile environments.

Nauru is an isolated raised coral-limestone atoll island [17] standing 4300 m above the ocean floor and located 41 km south of the Equator in the Pacific Ocean (Figure 2).

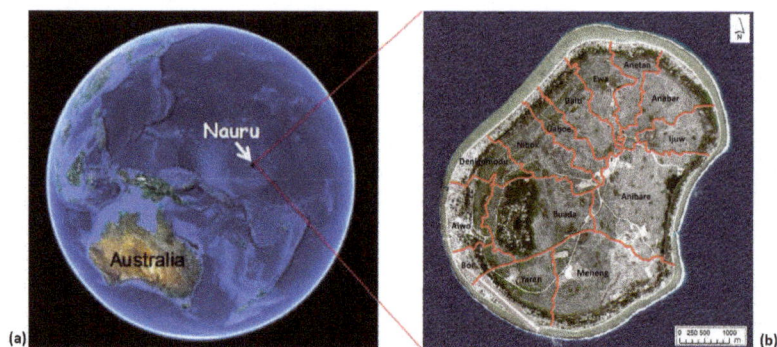

**Figure 2.** Nauru position at 0°32′ S and 166°56′ E (**a**); and aerial photo of the island in 2009 with its districts (**b**).

In 2001, the World Health Organization (WHO) [18], collaborating with Nauru public authorities, formulated a Long-Term Water Plan which has to be considered the starting point of any future water resources-related action. The main potential supply options identified by WHO for Nauru, also resumed in the National Sustainable Development Strategy 2005–2025 of Nauru [19], are: (a) extraction of freshwater from the shallow part of the aquifer; (b) collection and storage of rainwater; and (c) additional desalination plants. Point (c) looks to be the hardest one to face, since it is strictly linked to the problem of additional power energy supply based on renewable energy instead of fuel. Until 2015, an effective national integrated water resource management plan for Nauru was missing [20]. Then Nauru government published the Nauru Water and Sanitation Master Plan [21]. The plan is proposing as main supply options: (a) the collection and storage of rainwater; (b) a limited use of groundwater; and (c) building a reticulation network linking each house to desalinated water. The suggested limited use on groundwater was mainly due to the lack of knowledge of groundwater resources availability on the island.

Politecnico di Milano has been involved in a project focused on a feasibility study for the development of infrastructures for sustainable use of groundwater resources in Nauru [22]. The project, funded by Milano Municipality and related to EXPO 2015, aimed to seize the opportunity to enable the island to take an additional step toward a sustainable water supply, starting from the WHO results/directives [18–20] and providing Nauru with a new and improved system for groundwater exploitation. The characterization and modeling approach described in next pages could be followed by Pacific, Caribbean and Mediterranean islands that suffer the same problem. The entire project consisted of three steps: (1) development of the conceptual site model through the geological and hydrogeological characterization of the island; (2) implementation of 2D and 3D density-dependent flow and transport groundwater numerical models on the part of the island, which turned out to be more suitable for the groundwater development; and (3) study and preliminary design of infrastructural actions for groundwater sustainable exploitation. This paper focuses on the first step of the project and on 2D model results analysis that have made possible to clarify the mechanism for storing freshwater next to the seashore.

## 1.1. Nauru: Overview of the Island

Nauru has an oval shape, with an extension of about 22 km$^2$, and its 30 km of coastline are surrounded by a fringing coral reef 120–300 m wide (Figure 2). It is an independent state with a 2016 population of 9591 inhabitants. In the early 1980s, the island experienced a long economic wellbeing period thanks to the mining of the tricalcic phosphate deposits that originate from the droppings of sea birds. However, intensive phosphate mining has made 80 percent of Nauru's land very unusable (Figures S1–S3 in Supplementary Materials).

The climate of Nauru is hot and humid with mean daily temperatures of 29–31 °C and a mean minimum daily temperatures of 24–26 °C [23]. Annual rainfall data from 1946 show a mean annual rainfall of about 2100 mm, with a high variability (Figure S4 in Supplementary Materials). The wet season usually stretches from December to April, although El Niño Southern Oscillation (ENSO) events have a very strong impact on precipitation: a negative Southern Oscillation Index (SOI) is in high correlation with periods of high rainfall (El Niño period); conversely a positive SOI indicates periods of drought (La Niña period) as reported by Falkland [24]. El Niño phenomenon usually lasts for about 18 months and occurs on average every four years, although it is not completely predictable.

The total water for Nauru demand was estimated to be 1500 m$^3$/day of potable water and 1000 m$^3$/day of non-potable water. Today, potable water is supplied through the operation of the Nauru Phosphate Company (NPC) desalination plant in conjunction with the power station. This plant supplies 950 t/day of high quality potable water (1998/2001 data). Additional potable water is captured by houses through rainwater harvesting. The present supply of potable water can meet the population demand in rainy years, but not in dry years or when the plant runs low due to technical issues or the high oil prices.

The main concentration of groundwater wells is located along the coastline and around Buada Lagoon, where pumping or bailing are the methods used for water extraction. Groundwater is used in the island, even though in most wells it does not meet WHO drinking water standards set at 1000 mg/L of Total Dissolved Solids (TDS) [25]: consumption is mostly related to flushing toilets, showers and house work. However, during drought periods, it is used for wide purposes even in those areas where the TDS content is very high (e.g., for cooking/boiling and animals breeding). Acceptability of groundwater quality may vary not only according to the season but also depending on local circumstances: water tanks and rainwater harvesting systems maintenance, delays in delivery of desalinated water and people awareness about water quality/risks. For this reason, Jacobson et al. [26] suggested to set an upper limit for drinking water at 1500 mg/L of TDS.

## 1.2. Previous Hydrogeological Studies

Several hydrogeological studies on Nauru Island have been conducted in the past but few of them are reported in scientific publications. Jacobson et al. [26] have previously carried out hydrogeological studies of Nauru Island during the 1980s. The studies included core drilling, water electrical conductivity (EC) measurements inside drilled boreholes, and electrical sounding for determination of resistivity profiling on the island. The studies resulted in the construction of a conceptual model of the underlying geological units of the island and the distribution of groundwater with the location of freshwater/saltwater interface. From the analysis of core drilling samples [27], the underlying rock resulted to be made up by limestone intensely karstified up to at least 55 m below sea level. The water table had an average level of 0.3 m above the sea level, with water flowing radially towards the sea.

The freshwater was evaluated to lay above a 60 m thick transition zone. The hypothesis made during these studies is that the thickness of mixed water may be due to high hydraulic conductivity of the limestone present in Nauru subsoil. As known, an important contribution to this high conductivity is the presence of karstic systems, which allow free and quick movement of seawater inland, behaving as a source of saltwater [28]. Another factor responsible for the wide transition zone was detected in the tidal fluctuations, which ranged from 0.33 m below mean sea level to 0.51 m above sea level [27]. A reversal of hydraulic gradient at the shoreline was identified, with drainage outwards at low tide, and seawater flow inwards at high tide. The tidal effects looked to have a significant influence on groundwater levels too, being about half of the amplitude of the ocean tidal stage throughout the island. Unlike the rest of the island, the Buada Lagoon levels had shown a lack of tidal response suggesting the idea that the lagoon is an independent system in which the lowering water levels could be potentially induced by evapotranspiration in drought periods [26]. Jacobson and Hill [27] in 1987 did a groundwater EC survey for determining the position of freshwater lenses in the island subsoil. Two large freshwater lenses (with thickness greater than 7 m) were firstly identified [27], one located close to the Buada Lagoon depression, covering approximately 2.4 km², and the other on the north-central part of the island, extending for about 1.3 km². A threshold concentration of 1500 mg/L TDS was used for the localization and estimation of the freshwater lenses thickness [29]. However, the freshwater lenses located in the Topside area seemed not to be resilient: by contrast to the first investigations, in 2008, Falkland [24] performed a groundwater EC survey, after an extended dry period, showing very limited fresh groundwater resources, except for some boreholes by the northern coastal fringe in Ewa and Anetan districts, suggesting the possible presence in that area of a small freshwater lens resilient to sweater intrusion even in periods of droughts.

One of the most recent documents regarding groundwater assessment in the island is by Bouchet and Sinclair [30]: an EC survey was taken in 283 domestic wells on the coastal fringe of the island in the period March–April 2010. Considering the water of those wells fresh with a value of electrical conductivity smaller than 2500 µs/cm, the study reported an average freshwater thickness in the coastal fringe of about 0.8 m. These measures were taken at the end of a period of 8 months with rainfall above the average value. This suggested a quite significant impact of rainfall on groundwater salinity and a quick response of the system to rainfall changes. The 2008 and 2010 surveys [30] show that groundwater in Nauru is highly sensitive to climate variability, and therefore highly vulnerable, as already underlined by White and Falkland [7,31]. Nevertheless, the reason of the presence of these lenses was not clarified and the freshwater amount stored underground was not quantified.

## 2. Materials and Methods

This paper takes the main information from the activities completed by the authors during the project developed by Politecnico di Milano. This project (2010–2015) aimed to start the exploitation of groundwater resources addressing the extraction from the fresh surface layer of groundwater. For this reason, the study has been carried out to investigate the current situation of seawater intrusion in the Nauru aquifer. Starting from the data available in previous studies, in this work, the hypothesis arose that freshwater accumulates in the sandy part of the island, close to the coastline. The intent of the project was then to further the conceptual model of the island and verify the hypothesis of freshwater storage, in order to evaluate where more resilient fresh groundwater lenses are present and accordingly address groundwater extraction in the more feasible zones. Politecnico di Milano has carried out the field activities in collaboration with the Nauru Rehabilitation Company (NRC) and Ministry of Commerce, Industry and Environment (CIE) technicians. The results reached in the different steps of this study are deeply described in the documents [32–35] available on the Nauru project website [22].

### 2.1. Island Characterization

#### 2.1.1. Geomorphology and Geology

The land area of Nauru consists in a narrow coastal plain, which encircles a limestone escarpment. From a morphological point of view, the island is divided in three parts: Topside, Bottomside and Fringing Reef (Figure 3). The lowest zone (Bottomside) is the coastal plain that presents a maximum width of 400 m and with an elevation ranging from 0 to 10 m a.s.l. The Bottomside consists of a sandy or rocky beach on the seaward edge and a beach ridge or foredune, behind which are either relatively flat ground or, in some places, low-lying small lagoons, filled by brackish water (Ijuw and Anibare districts, Figure 2). Scattered limestone outcrops or pinnacles are found on both the coastal plain and on the intertidal flats of the fringing reef, with particularly good examples in the Anibare Bay area. The Fringing Reef encircles the whole island and extends from the coastline for about 150 m; the maximum width, about 250 m, is reached in the northern part, close to the Ewa and Anetan districts.

**Figure 3.** Reef, Bottomside and Topside of Nauru and location of wells and monitoring wells present in the island.

The transition between the Bottomside and the Topside is a scarp about 30 m high. Relative elevations on the plateau generally vary between 20 and 45 m a.s.l.; the highest point on the island, Command Ridge, is located in the west and presents an elevation of 70 m a.s.l. The outstanding topography after completion of primary phosphate mining is a pinnacle and pit relief.

On the southwest-central part of the island, there is a wide and fertile depression (about 120,000 m$^2$), where a brackish water lake known as Buada Lagoon is located. This is the only surface water body (38,000 m$^2$) existing on the entire island. Another depression, Ewa depression, is located in the northern sector (Figure 4) and extends for about 200,000 m$^2$. Unlike the Buada Lagoon, the Ewa depression is dry, although we might formulate the hypothesis that also there a lagoon was present in the past.

The first phase of the characterization was the Digital Terrain Model (DTM) implementation from the photogrammetric survey carried out by NRC in 2009 and elaborated by Politecnico di Milano during the first months of 2010. Using the developed DTM, it was possible to distinguish the low-lying zones from the most elevated ones (Figure 4 and Figure S5 in Supplementary Materials) with the aim of evaluating preferred runoff zones or areas where rainwater can accumulate. In a preliminary phase, this also allowed knowing the elevation of the stratigraphic logs and wells available in the area before the topographic survey was carried out.

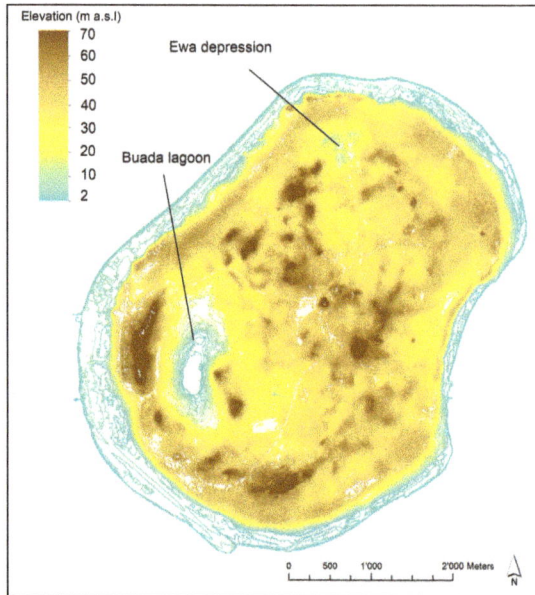

**Figure 4.** DTM elaborated by Politecnico di Milano with the data of the photogrammetric aerial survey carried out in 2010. Bottomside, cyan; and Topside, brown.

The information related to the subsoil has been drawn from the 127 geo-referenced wells present in the island (Figure 3). Some of these points are now used uniquely for groundwater level and electrical conductivity monitoring (35), while others (22) have been realized for the previous studies [26] but are nowadays impossible to be tacked down or used. The rest are private wells used for domestic purposes and rarely monitored by public authorities.

From the subsoil data, it has been possible to define the geological setting of the island. The whole Bottomside consists of coastal homogeneous sandy sediments that are present until about 5 m from the ground. Below these sediments, there are the carbonate rocks, as documented by Jacobson et al. [26], that distinguished the pliocenic dolomitic limestone located on the Topside from the pleisto-holocenic limestone that constitute the reef. From a geological point of view, the Topside consists of a matrix of coral-limestone pinnacles and limestone/dolomitesoutcrops, between which lie extensive deposits of soil and high-grade tricalcic phosphate rock. The phosphate has been mined for the past century. Due to the mining activity, the topsoil layer of around 75% of the island has been exploited cutting the natural vegetation quite completely [36–38]. The karst phenomenon in Nauru is very diffuse and the limestone assumes Karren shapes. Ewa depression can be interpreted as a sinkhole that formed because of subsidence phenomena. The stratigraphic logs highlight the great number of voids and caves present in the subsoil and caused by the karstic phenomenon; this phenomenon had modified and altered the groundwater flow, creating caves that in some cases may be observed at the ground level (Moqua and Anetan caves). The stratigraphic log analysis and the conceptual site model developed in this study lead to the conclusion that, from a hydrogeological point of view, it is possible to identify two aquifers in the island: one is represented by the Bottomside (made of sandy sediments), the other is located in the limestone and dolomite that make up the rest of the island. These two aquifer systems are connected vertically and laterally, as there are no separating low permeability levels. Probably, water rapidly crosses through the Topside and then slows down in the sandy aquifer of the Bottomside before discharging into the sea.

2.1.2. Piezometric Survey—Nauru Datum and Tidal Correction

To obtain a confirmation of the definitive conceptual site model and get data useful for the successive calibration of the numerical model, different piezometric surveys have been carried out by Politecnico di Milano. The identification of the flow field in presence of low hydraulic gradient needs an accurate knowledge of the monitoring point's elevation. For this reason, in October 2011 Politecnico di Milano researchers performed a GNSS (Global Navigation Satellite System) survey and 35 monitoring wells plus seven private wells have been geo-referenced, using two double frequency GNSS receivers (Table S1 in Supplementary Materials). One of the two receivers has been used as "master station" and has been located in a fixed point close to the NRC office. The second receiver has been used as rover, estimating the position of the remaining points. The double-difference method (with the point surveyed with the master station) has been used to post-process the data. This method consists in the evaluation of the differences in measurements obtained by the two different GNSS receivers observing more than two different GNSS satellites at the same time. This technique allows the removal of systematic errors due to receivers and satellites; for short base length (as in Nauru case), it also allows to considerably reduce the atmospheric signal propagation errors [34]. After the post-processing of the GNSS observations, the ellipsoidal heights of all the points have been obtained with a formal variance smaller than 1 cm.

Jacobson and Hill [27], for the first time, conducted a hydrogeological survey in Nauru and adopted a Reduced Level (RL) as reference point for all their measurements. Even today, all the hydrogeological data collected in Nauru are referred to RL (Figure S6 in Supplementary Materials). Thus, the measurements performed during the present GNSS survey needed to be linked to RL. However, the tide data provided by the Australian Bureau of Meteorology are referred to the Nauru Island Datum (NID), which corresponds to the zero tide gauge, located 0.166 m above the RL [39]. For the present study, the tide data have been corrected to refer them to the RL (Figure S7 in Supplementary Materials).

The Australian Bureau of Meteorology records the hourly tidal variations in the Aiwo district. Nauru Island has a micro-tidal regime and the average amplitude of the tidal signal is currently of about 130 cm compared to the mean sea level. It is a bit higher than the value reported in their study by Jacobson e Hill [27]. In the 2001–2011 decade, a maximum value of 3.15 m has been recorded in 2001 and a minimum of 0.36 m in 2011, compared to the RL [39]. In this study, 14 monitoring wells have been used to observe the groundwater variations related to the tidal fluctuations (Figure 5).

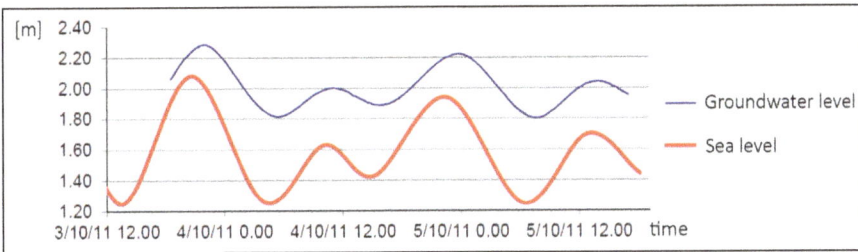

**Figure 5.** Sea and groundwater levels over a two-day observation period (compared to RL) in monitoring well S1 where the phreatic level varies within a range of 1.80–2.30 m every 6 h.

Falkland [24] had previously evaluated for the SIDS some parameters that characterize the groundwater variability in response to the tidal signal. Table 1 shows the average Tidal Lag and Efficiency of the tide calculated by Falkland in each monitoring well: the Tidal Lag (TL) expresses the delay between the occurrences of a tide maximum/minimum and the related maximum/minimum observed in the monitoring well. The Efficiency of the tide (E) is calculated as the ratio between the difference of maximum/minimum levels observed in the monitoring well and the related tidal difference ($\Delta G$ and $\Delta T$ in Figure 6).

**Table 1.** Tidal Lag, Efficiency and Distance from the coastline of the monitored monitoring wells.

| Monitoring Well | Location | Average Tidal Lag (hours:min) | Average Tidal Efficiency (E) | Minimum Sea Distance (m) |
|---|---|---|---|---|
| E2 | Topside | 01:37 | 0.50 | 1950 |
| E3 | Topside | 01:27 | 0.54 | 1500 |
| E8 | Topside | 01:49 | 0.49 | 950 |
| S3 | Topside | 01:55 | 0.47 | 820 |
| S9 | Topside | 01:32 | 0.46 | 520 |
| S10 | Topside | 01:51 | 0.44 | 1400 |
| S11 | Topside | 01:23 | 0.50 | 340 |
| S15 | Topside | 01:31 | 0.54 | 1040 |
| S21 | Topside | 01:20 | 0.41 | 397 |
| S22 | Topside | 01:20 | 0.49 | 290 |
| T1 | Topside | 01:27 | 0.56 | 1500 |
| **Average** | **Topside** | **01:33** | **0.49** | |
| S1 | Bottomside | 01:17 | 0.50 | 140 |
| S16 | Bottomside | 02:10 | 0.27 | 280 |
| T2 | Bottomside | 01:27 | 0.56 | 342 |
| **Average** | **Bottomside** | **01:38** | **0.44** | |

Implementation of a piezometric survey on the whole island needs about 5 h and this makes the measures influenced by the tidal event related to the monitoring time. In previous piezometric surveys, this issue has never been considered. Thus, here is proposed a method for data correction, developed during the project to adjust the measured levels by considering the tidal effect. Knowing the Tidal Lag (TL) and the tide Efficiency (E) for each measured monitoring well, it was possible to correct the measured groundwater levels referring them to the same time, as the measurements in all the monitoring wells have been done concurrently.

The correction process consists of four phases:

1. Reference time definition: A reference time ($t_0$) is fixed with the aim of referring all the field measurements to a standard time. Furthermore, $t_{obs}$ is defined as the real time corresponding to the instant when the hydraulic head is observed in the monitoring well.

2. Tide maximum variation definition: Analyzing the tide fluctuation curve it is possible to calculate the difference between minimum and maximum tide levels ($\Delta T$) that occur just before and after the groundwater head measurement. $\Delta t$ is the time lag between minimum and maximum tide level occurrences. Furthermore, the Tidal Lag (TL) can be used to evaluate the occurrence of the maximum or minimum level in the observed monitoring well. For the Nauru case, all tide data need to be corrected compared to the RL.

3. Correction factor estimation: The correction factor ($\Delta h_i$), which should be applied to the groundwater level measured at $t_{obs}$, can be calculated through the following equation:

$$\Delta h_i = E_i \frac{\Delta T}{\Delta t} |(t_0 - t_{obs})| \tag{1}$$

where $\Delta T/\Delta t$ represents the tide level variation velocity and $E_i$ is the efficiency for the monitoring well $i$. Consequently, ($E_i \Delta T/\Delta t$) is the groundwater level variation velocity due to the tide influence in the monitoring well considered.

4. Groundwater level correction: Four cases are possible, as summarized in Figure 6:

    i. The observation time ($t_{obs}$) and the reference time ($t_0$) are both between the same maximum and minimum groundwater levels and the head in the monitoring well ($h_{obs1}$) has been observed before the reference time $t_0$ (Figure 6a)—in this case, $\Delta h_i$ must to be subtracted from $h_{obs1}$.

ii.   The observation time ($t_{obs}$) and the reference time ($t_0$) are both between the same maximum and minimum groundwater level and the head in the monitoring well ($h_{obs2}$) has been observed after the reference time $t_0$ (Figure 6a)—in this case, $\Delta h_i$ must to be added to $h_{obs2}$.

iii.  The observation time ($t_{obs}$) falls before the maximum groundwater level but the reference time ($t_0$) is after the same maximum (see $h_{obs3}$ in Figure 6b)—in this case, the correction must account for the fact that from $t_{obs}$ to $t_0$ the groundwater level increases and then decreases. This means that $\Delta h_i$ must be added to $h_{obs3}$, but now it should be considered that $\Delta h_i = (\Delta h_{i,R} - \Delta h_{i,L})$ where $\Delta h_{i,R}$ is the correction during the groundwater rising phase and $\Delta h_{i,L}$ is the correction during the groundwater lowering phase.

iv.   The observation time ($t_{obs}$) falls after the minimum groundwater level but the reference time ($t_0$) is before the same minimum (see $h_{obs4}$ in Figure 6b)—in this case, the correction must account for the fact that from $t_0$ to $t_{obs}$ the groundwater level decreases and then increases. This means that the $\Delta h_i$ must be added to $h_{obs4}$, but now it should be considered that $\Delta h_i = (\Delta h_{i,L} - \Delta h_{i,R})$ where $\Delta h_{i,L}$ is the correction during the groundwater lowering phase and $\Delta h_{i,R}$ is the correction during the groundwater rising phase.

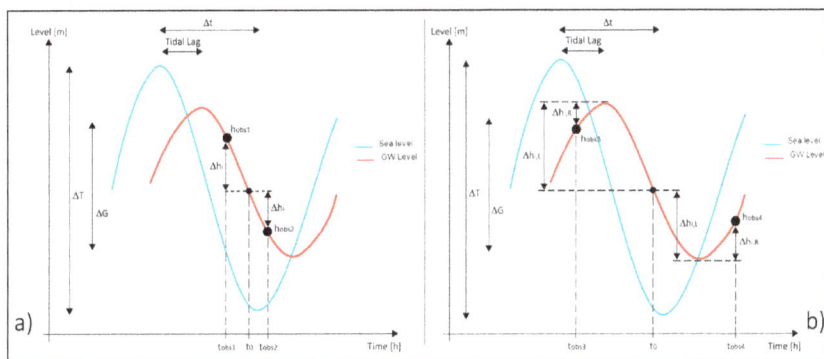

**Figure 6.** Groundwater level correction: Cases (i.) and (ii.) (**a**); and Cases (iii.) and (iv.) (**b**).

On the island, 23 piezometric surveys have been performed from June 2008 to April 2013 and interpolation after measures correction are showed in Results section.

### 2.1.3. Electric Conductivity Survey

Since 2008, during the piezometric surveys, the EC was measured at the surface and at the bottom of each monitoring well. In this way, the concentration distribution in plant and along the depth has been assessed. The monitoring wells in the southern part of the island show EC values always higher than 10,000 µS/cm. More in detail, the EC in the monitoring wells located in the northern sector along the coastline is around 520 µS/cm, while the internal monitoring wells (S11, S21, S23, S24 and T3) show an average value of 1300 µS/cm. The points located in the Topside, far from the northern sector, have higher values, often above the potable threshold value. In particular, close to the Buada Lagoon (E3, E7 and E8), the EC average values are around 11,300 µS/cm and the monitoring wells located in the airport area (Boe-Yaren districts) have elevated values of the parameter (around 5400 µS/cm). The monitoring well S3 always show very high values of EC (22,600 µS/cm), independently from the rainfall distribution. These high values can be justified by the fact that this zone had been used to deposit the sea sediments resulting from the building of the harbor facility at Anibar bay [40].

The data from the literature and the ones collected during this project have allowed drawing the following maps, where it is possible to see the salt concentration evolution into the Nauru aquifer from 2009 to 2013 (Figure 7 and Figure S8 and Table S2 in Supplementary Materials).

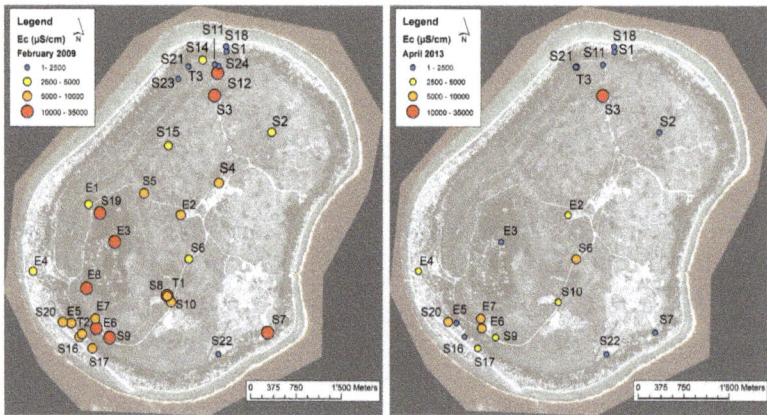

**Figure 7.** The 2009 and 2013 Electric Conductivity values at the water table into the Nauru aquifer.

## 2.2. Bi-Dimensional Numerical Model

To understand the cause of fresh water storage in the Bottomside area, a 2D steady state modeling phase was started, using the MODFLOW/SEAWAT finite difference codes [41,42]. The 2D model represents a preliminary study phase, preparatory to the hydrogeological parameter calibration for a following full 3D model implementation. For the bi-dimensional modeling, one section has been selected, the trace of which is shown in Figure 8. Section AA' has been chosen passing through the northern sector of the island, characterized by the thickest freshwater. The section has also been identified on the base of the position and number of monitoring wells and the head and concentration data available for the calibration process.

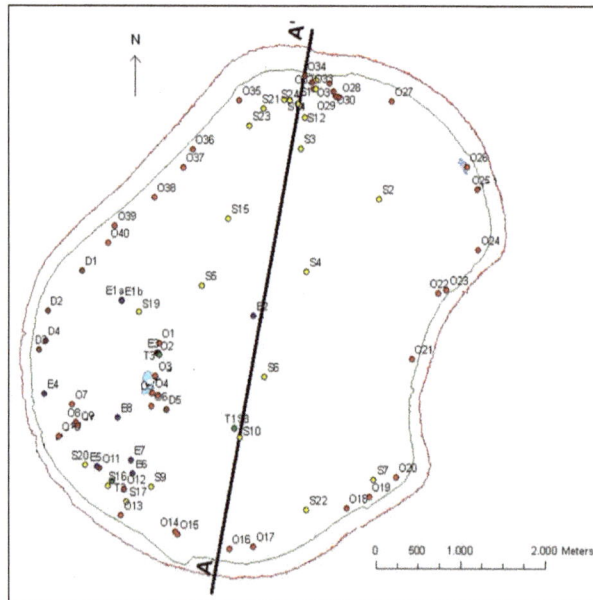

**Figure 8.** Section AA' trace and monitoring points present in the island.

Given that this is a problem of saltwater intrusion, the vertical discretization has been made very fine in the coastal zone where 1 m cells have been chosen (Figure 9). The cell size becomes bigger as we move inland, until a maximum size of 10 m.

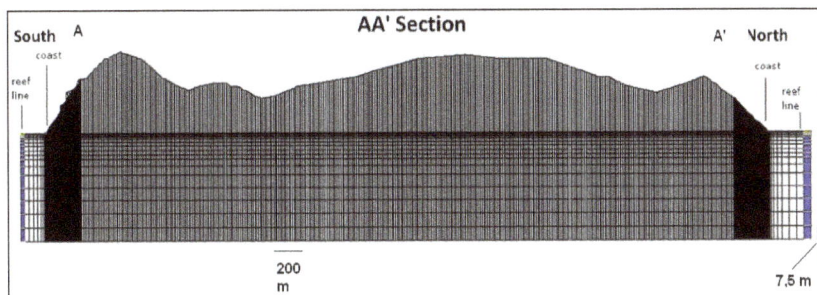

**Figure 9.** AA′ section, black zones represent the most refined part of the model domain (1 m cells).

For the vertical discretization, 18 layers have been set. The layer thickness increases from the surface to the bottom of the model (at −60 m respect to the RL) from a minimum value of 0.5 m for first four layers to a maximum of 7.5 for the last three. Constant head (1352 m above RL) and concentration ($35.7 \text{ kg/m}^3$ from Ghassemi et al. [29]) boundary conditions have been assigned at the borders of the two sections from the eighth layer to the last one. The layers above have been assigned boundary conditions to represent the sea bottom slope from the coastal line to the reef. Even though the hydrogeological investigation carried out in Nauru did not measure hydraulic conductivity, Jacobson and Hill [27] reported values for other limestone atoll islands where this parameter ranges between 1000 and 3500 m/day. For the assignment of parameters, the chosen initial values, uniform in the all model domain, have been the ones used by Ghassemi et al. [29] as shown in Table 2.

**Table 2.** Hydrogeological parameters initially assigned to the 2D models (from Ghassemi et al., [29]).

| Hydrogeological Parameter | Ghassemi Adopted Values | |
|---|---|---|
| Hydraulic conductivity (Kx, Kz) | 900 m/d | 18 m/d |
| Porosity | 0.3 | |
| Specific Storage and Specific Yield (Ss, Sy) | $0.0003 \text{ m}^{-1}$ | $0.3 \text{ m}^{-1}$ |
| Longitudinal dispersivity | 65 m | |
| Transverse dispersivity | 0.15 m | |
| Recharge | 540 mm/year | |
| Molecular diffusion | $8.64 \times 10^{-6} \text{ m}^2/\text{d}$ | |

The initial simulation has been run for a 200-year period, using an average recharge (540 mm/year) in order to simulate the natural extension of the saltwater wedge. The model was then calibrated in accordance to the head and concentration data relating to the 2010 survey carried out in wells and monitoring wells located along the section. The performed sensitivity analysis has shown that the model is highly influenced by horizontal hydraulic conductivity and dispersivity. As Ghassemi et al. [29] did not consider the presence of sand sediments along the coastline, the calibration process in this study has mainly considered monitoring wells S1 and S18 (Figure 8) located in that zone (Ewa and Anetan districts ), where the limestone is replaced by sand sediments in the shallow part of the aquifer. On these two points, the concentration values at different depths were available and have been used to compare the model results with real data. The calibration process followed a trial and error approach using 16 concentration data collected at different depths in eight multi-pipes monitoring wells. In Table 3 statistical results of the calibrated model are presented and the graphs in Figure 10 depict a detailed observed/simulated comparison along the depth in S1 and S18.

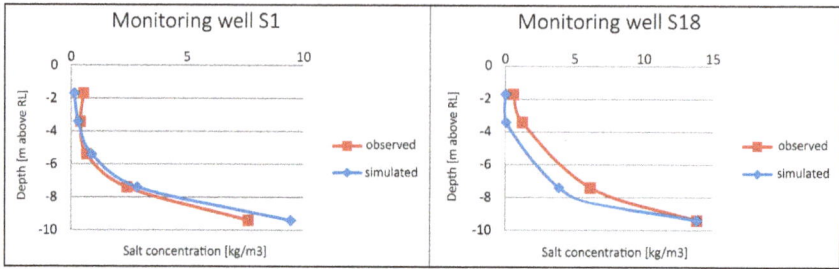

**Figure 10.** Observed vs simulated concentration values in monitoring wells S1 and S18.

**Table 3.** Statistical results of the salt concentrations (kg/m$^3$) calibrated with the 2D density-dependent model.

| Statistics | |
| --- | --- |
| Absolute Residual Mean | 1.70 |
| Residual Std. Deviation | 2.10 |
| RMS Error | 2.73 |
| Scaled RMS | 0.17 |
| Min. Residual | −0.28 |
| Max. Residual | 7.23 |
| Number of Observations | 16 |
| Range in Observations | 15.81 |

The calibrated hydraulic conductivity distribution adopted is the one shown in Figure 11 where the limestone zone (white part) is assigned $K_x = K_y = 800$ m/d, whereas the sand zone (red) is assigned $K_x = K_y = 40$ m/d (the ratio 0.1 is used for $K_z$).

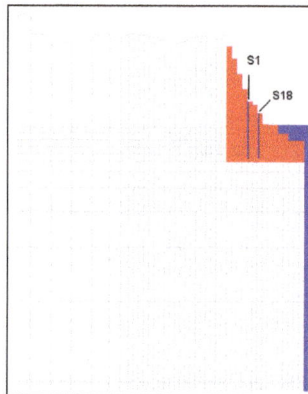

**Figure 11.** Calibrated hydraulic conductivity settings: sandy sediments (red) and limestone (white). In blue the constant head and constant concentration boundary conditions are indicated (zoom on the northern part of section AA′).

Even the dispersivity value has been the object of the calibration phase. In fact, tidal effects can produce considerable impacts on seawater intrusion processes in mixing zones. La Licata et al. [43] simulated seawater intrusion with and without tidal effects on a mixing zone; their results indicated that tidal mixing results in more mixed pollutant and salinity concentrations than the distributions from an equivalent steady-state model without tidal effects. For this reason, the dispersivity value has been slightly modified compared to Ghassemi et al. [29] on the limestone zone (50 m, 5 m and 0.2 m for longitudinal, transverse and vertical dispersivity, respectively), whereas it has been assigned a different value on the sandy coastal zones (2 m, 0.2 m and 0.008 m). The increase in the dispersivity values in the limestone zone was also necessary to represent the role of the karstic canals that characterize the internal part of the island. Campana and Fidelibus [44] used in their model values that ranged 2–100 m to represent a gypsum formation characterized by variously oriented fractures and karst features.

## 3. Results

### 3.1. Heads Distribution

The collected data related to the groundwater fluctuations in response to tidal variations have made possible the representation of the piezometric surface for Nauru Island as measures were carried out at the same time. The surveys have been carried out on an average of about 30 monitoring wells. The data have been corrected through the already described methodology and the piezometric maps have been created in order to understand the hydraulic gradient values and the main flow directions. The Jacobson et al. [26] piezometric map shows a radial groundwater flow from the center of the island toward the coastline. The eight surveys performed for this study in 2010–2013, conveniently corrected, have enabled us to reach some conclusions regarding the groundwater surface morphology (Figure 12). A high level persists during the time in the area where monitoring wells E3 and S19 (2.24 m, November 2010) are located, close to the Buada Lagoon (2.01 m, November 2010). Generally, in the Topside the hydraulic gradient is very low (<0.05%) and increases moving from the cliff toward the Bottomside (about 0.1%). Locally, the coastal area presents piezometric values similar to or a bit higher than the monitoring wells located in the internal part of the Topside. This phenomenon is more significant in the northern part of the island, especially in the Ewa and Anetan districts, where relevant freshwater presence has been detected. All the surveys carried out during this project confirm the radial flow toward the ocean. The Buada Lagoon does not seem to modify or influence the piezometric surface, but contrary to previous studies it does not represent a drainage area and shows levels that respond to tidal variations and are comparable to the surrounding groundwater ones.

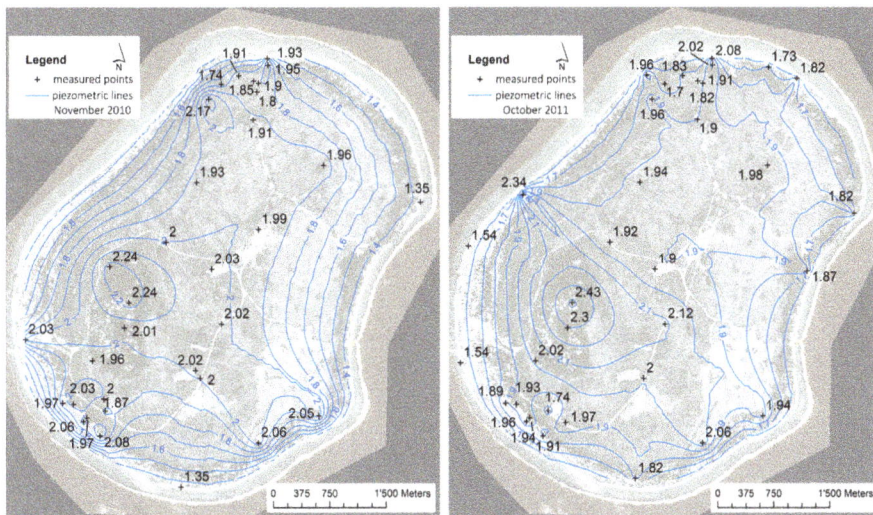

**Figure 12.** Piezometric maps of the groundwater heads measured in November 2010 and October 2011 (sea level 1.35 and 1.54 m, respectively).

## 3.2. Electric Conductivity Distribution and Exploitable Areas

Given that many monitoring wells are multi-pipe, it has been possible to evaluate the EC vertical profile and then the geometry of the saltwater wedge and the thickness of freshwater lenses. In the northern sector of the island, monitoring wells S1 and S18 show quite a high thickness of freshwater (about 7 m) that is also constant in time. The lens thickness decreases in the southern part where monitoring wells S21, S24 and S23 show a freshwater thickness of about 3.5 m, which remain stable only in S24, while in the other points are more influenced by the rain periods. The central sector of the Topside (S6, S10) does not show any freshwater availability given that the EC values are between 10,000 and 20,000 µS/cm at the surface and reach about 35,000 µS/cm at 20 m depth from the water table. In the area close to the airport, freshwater does not accumulate in large quantity and only S16 has a freshwater thickness of 3 m. This lens was present only in April 2010, which was very rainy (390 mm of rain), whereas, during the dry months, the freshwater previously stored in this zone flowed toward the sea. As described in the Introduction Section, previous studies [27] had determined two areas with a freshwater thickness higher than 5 m: one located close to the Buada Lagoon and the other near the S15, S4 and S2 monitoring wells. The hydrogeological surveys carried for this study combined with the Falkland ones, had shown that in these zones there is no resilient freshwater lens; this means that the areas determined by Jacobson and Hill are strictly linked to a very rainy period (1986–1987) and not to particular hydrogeological structures able to store freshwater. Figure 13 sums up these results. The characterization activities carried out during this study have made possible to determine the areas most suitable for the design and development of groundwater infrastructures for water withdrawal (e.g., wells and infiltration galleries). The most suitable area is surely the northern zone where S1 and S18 are located, given that there have been found freshwater lens with a thickness resilient in time and independent from the rainfall distribution. A quite thick freshwater lens (3 m) is also located in the southern part of the island, in Yaren and Boe districts.

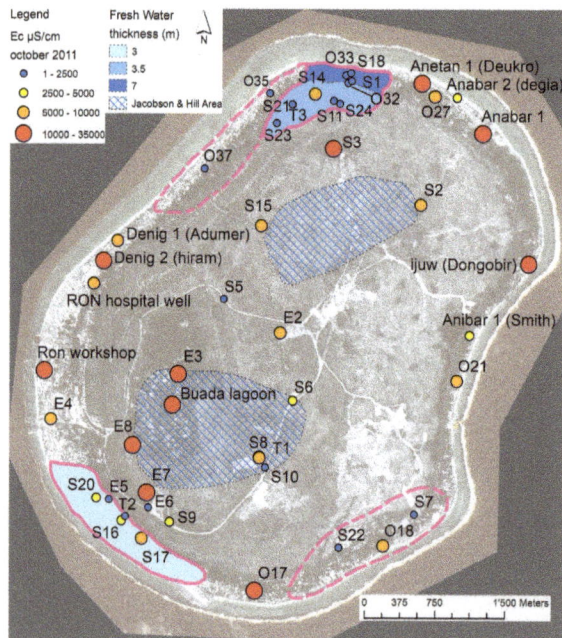

**Figure 13.** Results of the EC survey in monitoring wells and private wells (October 2011). Pink line indicates the most suitable areas for groundwater exploitation and the pink dashed one the potential suitable areas.

*3.3. Model Results*

As the surveys clearly show that the largest volume of freshwater is present in the northern part of the island, modeling activities focused on coastal monitoring wells (S1–S18) to understand the mechanism that allows this storage. Considering the calibration statistics and the S1 and S18 results (Figure 10), the model can be considered able to properly represent the salinity vertical distribution. In S1, located about 150 m from the sea, a freshwater thickness of 7 m is simulated with a following sharp increase in concentration values. In S18, located about 50 from the sea, the model results are a little bit worse because simulated concentrations are lower than the measured ones. The model properly represents the concentration at three of the four available depths, but it underestimates the salinity at the depth of −7.5 m above RL, consequently overestimating the freshwater thickness of about 25%. This difference is because S18 is very near the coast line and is strongly influenced by tide level variation. For the future 3D modeling phase, this aspect should be considered and, probably, daily average salinity values, rather than instant measurements, would be more suitable for a steady state simulation. In Figure 14, concentration values are visible along the northern portion of the cross section examined. The saltwater wedge from the cost toward the inland is clearly depicted.

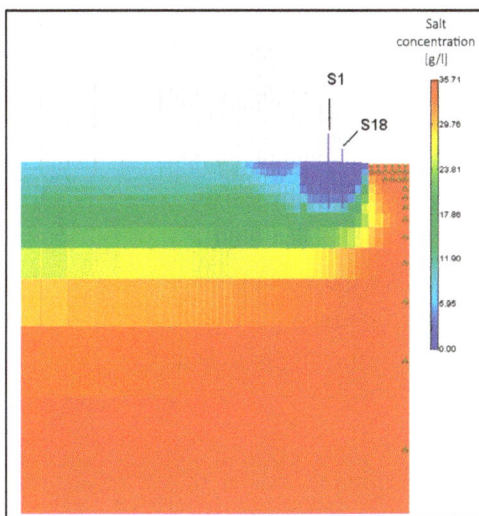

**Figure 14.** Concentration distribution of the calibrated model (kg/m$^3$) simulated for November 2010 (zoom on the northern part of section AA); green dots indicate cells hosting the boundary condition.

Thanks to the low hydraulic conductivity, freshwater is stored in sand sediments creating a sort of "pillow" that has a sharp limit toward the seaside and a smoothed one toward the inland side. The low hydraulic conductivity of the sands, slowing the groundwater flow, also results in a decrease of groundwater concentrations in the limestones area immediately upgradient of the S18–S1 area. Nevertheless, these freshwater lenses are thin and with concentration values above 4 kg/m$^3$ that limit the exploitation possibilities.

The 2D model results highlight the fact that the transition from limestone to sand occurring in the Bottomside is responsible for the freshwater storage mainly observed in the northern zone of the island. In particular, the study conducted and the calibration process of the mathematical models have highlighted the fact that the coastal zone, filled by sands, is the one where the biggest thickness of freshwater occurs. This is because the hydraulic conductivity of this area, smaller than the limestone internal zone, allows the flow to slow down and the groundwater to store. The 2D model demonstrated that even the dispersivity values and the recharge play a role in concentration distribution.

## 4. Discussion

The hydrogeological characterization and the modeling activity carried out, albeit preliminary, have enabled to achieve important new results in comprehending the freshwater lens formation phenomenon in small islands.

In the case of Nauru Island, the investigations performed have unexpectedly allowed identifying the presence of freshwater lenses hosted into the sandy sediments of the coastal zone, close to the seashore. The monitoring activity, carried out for a six-year period, has moreover highlighted that these lenses are resilient even in drought conditions. This result is in contrast with the previous studies that, on the other hand, had identified freshwater lenses into the limestone forming the internal part of the island. The difference in the results is due to: (1) previous authors had carried out just a single survey corresponding to a particularly rainy period; and (2) their conceptual site model did not consider the presence of sandy sediments along the coastline. Thus, thanks to the data collected during this study, a new conceptual site model has been developed for Nauru, one presuming that the low hydraulic conductivity of sand makes the groundwater slow down toward the coast consequently allowing freshwater storage where saltwater is instead expected to penetrate more easily into the aquifer. This preliminary conceptual site model has been verified through a steady state 2D numerical model along a N-S oriented section. During periods with great rainfall, water easily infiltrates the subsoil because of the paucity of vegetation and the karst phenomenon that characterize the limestone making up the internal part of the island. Freshwater then flows in the high conductivity limestones (800 m/d) radially moving toward the coast and slowing down when it reaches the low permeability sands (40 m/d). Here, freshwater stores in lenses that remain protected from saltwater intrusion by the low hydraulic conductivity of the sediments. Unlike the case of the Topside, characterized by high hydraulic conductivity and dispersivity, the infiltrated rain quickly mixes with the deeper saltwater, thereby determining a fast disappearance of freshwater lenses in the periods with scarce rainfall distribution.

Three hydrogeological/geomorphological factors can be considered responsible for the large storage of fresh water in Nauru's northern sector

- The wind: During the humid season (November/April), the winds blow from the West, while, in the dry season, they come from East. This means that the northern and southern coastal zone are not particularly exposed to the wave action, and consequently there is a smaller erosion of the sands [40].
- The morphology and petrography of the coastal zone: The coastal zone has a variable width, from 400 m close to the airport to a few meters close to Anabar Bay. Near the S1 and S18 monitoring wells, the Bottomside width is 180 m and the sediments mainly consist of by carbonate sands. The main difference is linked to the thickness of sediments: in S1 and S18 the sediments are 15 m thick. Other boreholes in the Bottomside had shown that the basement is around 6 m from the ground level in the eastern sector, 10 m close to the airport and only 3 m in the Anibar bay. The elevated thickness of the sandy sediments in the northern zone is probably the cause of the groundwater slowing down and of storage. Here, the flow circulation is therefore different from the other zones where instead, due to the presence of karst tunnels, the groundwater flows rapidly toward the sea.
- The bathymetry: Looking at the bathymetric maps of Nauru [45], it is possible to notice that in the northern sector the sea bottom declines smoothly compared, for example, to the Anabar Bay zone where submarine cliff is present. That probably allows the decrease of wave intensity in the northern sector, thereby determining the sand sedimentation and the freshwater storage in the S1 and S18 area.

Comprehension of the way freshwater lenses form in the subsoil of small islands and how they behave is a fundamental aspect to be understood in order to achieve a sustainable development of groundwater. The Nauru experience show how, on small islands, with the aim to archive a proper groundwater management, a detailed aquifer characterization and an exhaustive implementation of the hydrogeological conceptual site model are necessary. Nevertheless, even for Nauru, the studies presented in this paper can not be considered sufficient to prepare a reliable groundwater management plan. The complete understanding of freshwater storage and its quantification would need a more accurate 3D modeling phase supported by data collection. In particular it would be useful a series of hydraulic conductivity tests to confirm the parameters used in the 2D modeling and some geo-electrical tomography campaigns to better assess the freshwater thickness along the coast. This new investigations have been designed for Nauru and will indeed allow: (i) a more precise simulation of the system behavior; (ii) a better comprehension of the role played by the sand sediment thickness in the different zones of the island; and (iii) an improved assessment of the volume of stored freshwater and its resilience capacity during drought periods. Furthermore, an unsteady state 3D model will constitute a useful tool for management of water resources, their protection from pollution and definition of a sustainable use of freshwater lenses avoiding saltwater intrusion increasing. Through this kind of model, it will possible to design new and tailored groundwater withdrawal systems, forecasting their effects on groundwater availability and on saltwater intrusion in different meteorological scenarios. This will allow optimizing their number, position, depth and pumping helping public authorities in preparing a groundwater management plan in the aim to achieve a sustainable development.

## 5. Conclusions

On Pacific, Caribbean and Mediterranean small islands, groundwater is an important, or the main, source of freshwater whose availability is often limited and whose quality is frequently compromised. The present study of Nauru's hydrogeology is a step toward the knowledge of groundwater behavior into highly permeable aquifers, underlain and surrounded by seawater, and can contribute to improve the global knowledge on water management in these fragile systems. The study findings are mainly related to the following aspects:

- Unlike generally assumed, small island aquifers can not only host continuous freshwater lens in the central part of the island, but, unexpectedly, freshwater storage can also occur next to the coastline. In Nauru case, long-term investigations carried out by authors, have shown that those lenses are resilient to saltwater intrusion even in drought periods.
- A method to correct head measurements vs. tide has been proposed and applied to better characterize the groundwater flow patterns in small islands.
- Thanks to the investigations and the numerical modeling, it has been possible to clarify the mechanism for freshwater storage next to the seashore and the role played by the hydrogeological structure and aquifers hydraulic conductivity.
- In previous studies, the durability of freshwater lenses had not been proven yet; the characterization activities here presented cover a long period and show that freshwater lenses located along the coastline turn out to be resilient to drought and saltwater intrusion.

Despite the previous groundwater investigations, considering climate change, rising sea levels and increasing frequency of extreme events, there was a need for a more comprehensive assessment of groundwater potential on Nauru. The characterization presented in this study makes available new and more useful data related to the hydrogeological setting of the island. The results achieved have highlighted the existence of a fresh groundwater resource that can be exploited, in a sustainability perspective. The volume stored in the subsoil will have to be better quantified in the future both in the northern sector of the island and throughout the coast. Probably groundwater alone would not be sufficient to meet the needs of the Nauru population and therefore, as the island's Water Plan suggests, the use of the groundwater must undoubtedly be combined with others water resources. However, the use of this resource alongside rainwater harvesting is an important resource to ensure the island's future water security, even during periods of drought or desalination plant breakdown. Due to the vulnerability of fresh groundwater lenses, their use should be carefully managed in order to avoid any uncontrolled phenomena of saltwater intrusion and the overexploitation of the resource. Furthermore, at least where fresh groundwater is available, strict rules are necessary to avoid any pollution by anthropogenic activities (mainly cesspits/septic tanks and animals breeding). Groundwater is a natural resource that should be considered as public and shared resource for present and future generations, even more on small islands where are present some of the most vulnerable aquifer systems in the world. Fresh groundwater should not be freely exploited by any private entity without control, and it would be desirable that in the future the State would directly assume the responsibility for extracting and distributing water. Consequently, Nauru's next Water Plan would have to addresses these issues, ruling groundwater exploitation and pollutant activities, as well as adopting a program of groundwater survey and monitoring wells maintenance.

**Supplementary Materials:** The following are available online at www.mdpi.com/2073-4441/9/10/788/s1, Figure S1: Nauru Topside—landscape, 2010; Figure S2: Nauru Topside—Scattered limestone outcrops (or pinnacles), 2010; Figure S3: Nauru Bottomside, 2010; Figure S4: Annual rainfall in Nauru from 1946 to 2015 (data from Nauru Government and Australian Bureau of Meteorology [46]); Figure S5: 3D visualization of the DTM elaborated by Politecnico di Milano with the data of the photogrammetric aerial survey carried out in 2010; Figure S6: Mean sea level and relationship between Reduced Level (or Nauru Datum) and tidal height (by Jacobson et al. [26]); Figure S7: Pattern of the leveling operations in order to obtain the orthometric elevation of RL from H1. The Australian Bureau of Meteorology uses the Nauru Island Datum (NID) as reference level for the tide measurement; Figure S8: Salt concentration evolution into the Nauru aquifer from 2008 to 2013 at the water table, Table S1: Coordinates of the points directly observed trough GNSS receiver. RL stays for the elevation measured above the Reduced Level. Annex 1: Monographs of the surveyed monitoring wells. Table S2: Electrical Conductivity [µS/cm] at water table (data referred to Figure 7 in the paper and S8).

**Acknowledgments:** This study was supported by the Nauru Project (2010–2015), which was funded by Milano Municipality and related to EXPO 2015. Authors did not received funds for covering the costs to publish in open access. The authors thank the Nauru Rehabilitation Company (NRC) and the technicians of Ministry of Commerce, Industry and Environment (CIE), for carrying out the field activities in Nauru Island in collaboration with the researchers of Politecnico di Milano. The authors are also grateful to Tony Falkland for being available in discussing groundwater data, groundwater investigations and possible groundwater development options for Nauru. The authors would like to thank Daniel Feinstein (U.S. Geological Survey—Wisconsin) for his help and his assistance in the model implementation and run. Finally, the authors would like to thank Louis Bouchet for his help in the field and for being the connection between the researchers of Politecnico di Milano and Nauru Island.

**Author Contributions:** The authors have contributed to this study carrying out different activities including project organization and development, characterization surveys, model implementation and calibration. All authors conceived and designed the methodology and discussed simulation results. Luca Alberti and Ivana La Licata, as experts in the hydrogeology area and in groundwater numerical modeling through MODFLOW/SEAWAT codes, conceived and designed the project for Nauru Island. Luca Alberti and Martino Cantone performed the survey activities on Nauru Island and analyzed the data. Martino Cantone, who is an expert in hydrogeology area and skilled in mapping tools, carried out the data organization. Luca Alberti and Ivana La Licata implemented the 2D numerical model. All authors carried out the calibration process and discussed model results. Therefore, the manuscript was written by the three authors in equal parts.

**Conflicts of Interest:** The authors declare no conflict of interest.

## References

1. White, I.; Falkland, T. Reducing groundwater vulnerability in Carbonate Island countries in the Pacific. In *Climate Change Effects on Groundwater Resources: A Global Synthesis of Findings and Recommendations*; Gurdak, J.J., Ed.; CRC Press: Boca Raton, FL, USA, 2011; Volume 27, pp. 75–110. ISBN 978-0-203-12076-7.
2. South Pacific Applied Geoscience Commission. *ADB Pacific Regional Action Plan on Sustainable Water Management*; South Pacific Applied Geoscience Commission: Suva, Fiji, 2003.
3. Falkland, T. From Vision to Action Towards Sustainable Water Management in the Pacific. In *Theme 1 Overview Report, Water Resources Management*; Ecowise Environmental: Camberra, Australia, 2002.
4. White, I.; Falkland, T.; Perez, P.; Dray, A.; Metutera, T.; Metai, E.; Overmars, M. Challenges in freshwater management in low coral atolls. *J. Clean. Prod.* **2007**, *15*, 1522–1528. [CrossRef]
5. Werner, A.D.; Sharp, H.K.; Galvis, S.C.; Post, V.E.A.; Sinclair, P. Hydrogeology and management of freshwater lenses on atoll islands: Review of current knowledge and research needs. *J. Hydrol.* **2017**, *551*, 819–844. [CrossRef]

6.    Falkland, T. Water resources issues of small island developing states. *Nat. Resour. Forum* **1999**, *23*, 245–260. [CrossRef]

7.    White, I.; Falkland, T. Practical Responses to climate change: Developing National Water Policy and Implementation Plans for Pacific Small Island Countries. In *Water and Climate: Policy Implementation Challenges, Proceedings of the 2nd Practical Responses to Climate Change Conference, Canberra Australia, 1–3 May 2012*; Engineers Australia: Canberra, Australia, 2012; pp. 439–449.

8.    Chen, Z.; Grasby, S.E.; Osadetz, K.G. Relation between climate variability and groundwater levels in the upper carbonate aquifer, southern Manitoba, Canada. *J. Hydrol.* **2004**, *290*, 43–62. [CrossRef]

9.    Ma, T.; Wang, Y.; Guo, Q. Response of carbonate aquifer to climate change in northern China: A case study at the Shentou karst springs. *J. Hydrol.* **2004**, *297*, 274–284. [CrossRef]

10.   Gattinoni, P.; Francani, V. Depletion risk assessment of the Nossana Spring (Bergamo, Italy) based on the stochastic modeling of recharge. *Hydrogeol. J.* **2010**, *18*, 325–337. [CrossRef]

11.   Scott, D.; Overmars, M.; Falkland, T.; Carpenter, C. *Pacific Dialogue on Water and Climate*; South Pacific Applied Geoscience Commission: Suva, Fiji, 2003.

12.   Ayers, J.F.; Vacher, H.L. Hydrogeology of an Atoll Island: A Conceptual Model from Detailed Study of a Micronesian Example. *Groundwater* **1986**, *24*, 185–198. [CrossRef]

13.   Bailey, R.T.; Jenson, J.W.; Olsen, A.E. Numerical modeling of Atoll Island hydrogeology. *Groundwater* **2009**, *47*, 184–196. [CrossRef] [PubMed]

14.   Nakada, S.; Umezawa, Y.; Taniguchi, M.; Yamano, H. Groundwater Dynamics of Fongafale Islet, Funafuti Atoll, Tuvalu. *Groundwater* **2012**, *50*, 639–644. [CrossRef] [PubMed]

15.   Vacher, H.L. Introduction: Varieties of carbonate islands and a historical perspective. In *Geology and Hydrogeology of Carbonate Islands*; Vacher, H.L., Quinn, T.M., Eds.; Elsevier: Amsterdam, The Netherlands, 1997; pp. 1–33. ISBN 0444815201.

16.   Adger, W.N. Social-Ecological Resilience to Coastal Disasters. *Science* **2005**, *309*, 1036–1039. [CrossRef] [PubMed]

17.   Hill, P.J.; Jacobson, G. Structure and evolution of Nauru Island, central Pacific Ocean. *Aust. J. Earth Sci.* **1989**, *36*, 365–381. [CrossRef]

18.   Wallis, I. *Draft National Water Plan for Government of Nauru*; Prepared in Cooperation with Ministry of Health; Workshop Notes September 2001, in press.

19.   Government of the Republic of Nauru. *National Sustainable Development Strategy*; Ministry of Finance and Economic Planning, Development Planning and Policy Division: Yaren, Nauru, 2005.

20.   South Pacific Applied Geoscience Commission. *Sustainable Integrated Water Resources and Wastewater Management in Pacific Island Countries National IWRM Diagnostic Report, Nauru*; South Pacific Applied Geoscience Commission: Suva, Fiji, 2007.

21.   Government of the Republic of Nauru. *Nauru Water and Sanitation Master Plan*; European Union: Brussels, Belgium, 2015.

22.   Nauru Project. Available online: http://nauru.como.polimi.it (accessed on 15 September 2017).

23.   Australian Bureau of Meteorology; CSIRO. *Climate Change in the Pacific: Scientific Assessment and New Research. Volume 1: Regional Overview*; Australian Bureau of Meteorology: Melbourne, Australia; CSIRO: Canberra, Australia, 2011; Volume 2.

24.   Falkland, T. *Country Implementation Plan for Improving Water Security in the Republic of Nauru*; South Pacific Applied Geoscience Commission: Suva, Fiji, 2010.

25.   World Health Organization. *Total Dissolved Solids in Drinking-Water Background Document for Development of Health Criteria and other Supporting Information*; World Health Organization: Geneva, Switzerland, 1996.

26.   Jacobson, G.; Hill, P.J.; Ghassemi, F. Geology and Hydrogeology of Nauru Island. In *Geology and Hydrogeology of Carbonate Islands*; Vacher, H.L., Quinn, T.M., Eds.; Elsevier: Amsterdam, The Netherlands, 1997; pp. 707–742. ISBN 0444815201.

27.   Jacobson, G.; Hill, P.J. Hydogeology and groundwater resources of Nauru Island, Central Pacific Ocean. *Groundwater* **1988**, *12*, 85.

28.   Romanazzi, A.; Gentile, F.; Polemio, M. Modelling and management of a Mediterranean karstic coastal aquifer under the effects of seawater intrusion and climate change. *Environ. Earth Sci.* **2015**, *74*, 115–128. [CrossRef]

29. Ghassemi, F.; Jakeman, A.J.; Jacobson, G.; Howard, K.W.F. Simulation of seawater intrusion with 2D and 3D models: Nauru Island case study. *Hydrogeol. J.* **1996**, *4*, 4–22. [CrossRef]

30. Bouchet, L.; Sinclair, P. *Assessing the Vulnerability of Shallow Domestic Wells in Nauru*; Technical Report 435; South Pacific Applied Geoscience Commission: Suva, Fiji, 2010.

31. White, I.; Falkland, T. Management of freshwater lenses on small Pacific islands. *Hydrogeol. J.* **2010**, *18*, 227–246. [CrossRef]

32. Alberti, L.; Cantone, M.; La Licata, I. Carried Out Activities in Nauru 24 November–1 December 2010. Available online: http://nauru.como.polimi.it/activities-report-november-2010/activities-report-november-2010 (accessed on 15 September 2017).

33. Alberti, L.; Cantone, M.; La Licata, I.; Oberto, G. Carried Out Activities in Nauru 29 September–13 October 2011. Available online: http://nauru.como.polimi.it/activities-report-november-2010/activities-report-october-2011 (accessed on 15 September 2017).

34. Alberti, L.; Cantone, M.; Oberto, G.; Sampietro, D. GNSS Static Suvey Report. Available online: http://nauru.como.polimi.it/activities-report-november-2010/gnss-nauru-survey-report-oct-2011 (accessed on 15 September 2017).

35. Alberti, L.; La Licata, I.; Cantone, M. PROGETTO NAURU—Sintesi delle attività del primo anno. Available online: http://nauru.como.polimi.it/activities-report-november-2010/nauru-report-july-2011 (accessed on 15 September 2017).

36. Dupon, J.F.; Bonvallot, J.; Florence, J. *Pacific Phosphate Island Environments Versus the Mining Industry: An Unequal Struggle*; South Pacific Regional Environment Programme, South Pacific Commission: Noumea CEDEX, New Caledonia, 1989.

37. Barrett, P.J. *Report on Phosphate, other Minerals and Groundwater Resources, and on Aspects of Rehabilitation Planning and Methodology, Nauru, Pacific Ocean*; Commission of Inquiry into the Rehabilitation of Mined-out Phosphate Lands of Nauru: Menen, Nauru, 1988.

38. Morrison, R.J.; Manner, H.I. Pre-mining pattern of soils on Nauru, Central Pacific. *Pac. Sci.* **2005**, *59*, 523–540. [CrossRef]

39. Australia Geoscience. *EDM Height Traversing Levelling Survey, Nauru*; Australia Geoscience: Canberra, Australia, 2009.

40. Maharaj, R.J. *Evaluation of the Impacts of Harbour Engineering, Anibare Bay, Republic of Nauru (RON)*; Technical Report 316; South Pacific Applied Geoscience Commission: Suva, Fiji, 2011.

41. Harbaugh, A.W. *MODFLOW-2005, The U.S. Geological Survey Modular Ground-Water Model—The Ground-Water Flow Process*; U.S. Department of the Interior: Washington, DC, USA, 2005; p. 253.

42. Langevin, C.D.; Guo, W. MODFLOW/MT3DMS-based simulation of variable-density ground water flow and transport. *Groundwater* **2006**, *44*, 339–351. [CrossRef] [PubMed]

43. La Licata, I.; Langevin, C.D.; Dausman, A.M.; Alberti, L. Effect of tidal fluctuations on transient dispersion of simulated contaminant concentrations in coastal aquifers. *Hydrogeol. J.* **2011**, *19*, 1313–1322. [CrossRef]

44. Campana, C.; Fidelibus, M.D. Reactive-transport modelling of gypsum dissolution in a coastal karst aquifer in Puglia, southern Italy. *Hydrogeol. J.* **2015**, *23*, 1381–1398. [CrossRef]

45. Kruger, J.; Sharma, A. *Nauru Technical Report. High-Resolution Bathymetric Survey, Fieldwork Undertaken on 30 September 2005*; EU EDF 8/9—SOPAC Project Report 116. Reducing Vulnerability of Pacific ACP States; South Pacific Applied Geoscience Commission: Suva, Fiji, 2008.

46. Australian Bureau of Meteorology. Available online: http://www.abs.gov.au/websitedbs/d3310114.nsf/home/Consumer+Price+Index+Inflation+Calculator (accessed on 15 September 2017).

*water*

MDPI

*Article*

# Groundwater Overexploitation and Seawater Intrusion in Coastal Areas of Arid and Semi-Arid Regions

**Nawal Alfarrah [1,2,\*] and Kristine Walraevens [1]**

[1]   Laboratory for Applied Geology and Hydrogeology, Department of Geology, Ghent University,
     Krijgslaan 281 S8, 9000 Ghent, Belgium; kristine.walraevens@ugent.be
[2]   Geology Department, Az Zawiyah University, Az Zawiyah, Libya
\*   Correspondence: nawalr2003@yahoo.com

Received: 19 December 2017; Accepted: 25 January 2018; Published: 2 February 2018

**Abstract:** The exploitation of groundwater resources is of high importance and has become very crucial in the last decades, especially in coastal areas of arid and semi-arid regions. The coastal aquifers in these regions are particularly at risk due to intrusion of salty marine water. One example is the case of Tripoli city at the Mediterranean coast of Jifarah Plain, North West Libya. Libya has experienced progressive seawater intrusion in the coastal aquifers since the 1930s because of its ever increasing water demand from underground water resources. Tripoli city is a typical area where the contamination of the aquifer in the form of saltwater intrusion is very developed. Sixty-four groundwater samples were collected from the study area and analyzed for certain parameters that indicate salinization and pollution of the aquifer. The results demonstrate high values of the parameters Electrical Conductivity, $Na^+$, $K^+$, $Mg^{2+}$, $Cl^-$ and $SO_4^{2-}$, which can be attributed to seawater intrusion, where $Cl^-$ is the major pollutant of the aquifer. The water types according to the Stuyfzand groundwater classification are mostly CaCl, NaCl and Ca/MgMix. These water types indicate that groundwater chemistry is changed by cation exchange reactions during the mixing process between freshwater and seawater. The intensive extraction of groundwater from the aquifer reduces freshwater outflow to the sea, creates drawdown cones and lowering of the water table to as much as 25 m below mean sea level. Irrigation with nitrogen fertilizers and domestic sewage and movement of contaminants in areas of high hydraulic gradients within the drawdown cones probably are responsible for the high $NO_3^-$ concentration in the region.

**Keywords:** seawater intrusion; coastal aquifer; arid and semi-arid regions; cation exchange; Tripoli; Libya

## 1. Overview Saltwater Intrusion into Coastal Aquifers

Coastal aquifers serve as major sources for freshwater supply in many countries around the world, especially in the Mediterranean [1]. The fact that coastal zones contain some of the most densely populated areas in the world makes the need for freshwater even more acute [1]. The intensive extraction of groundwater from coastal aquifers reduces freshwater outflow to the sea and creates local water table depression, causing seawater to migrate inland and rising toward the wells [2–4], resulting in deterioration in groundwater quality. This phenomenon, called seawater intrusion, has become one of the major constraints imposed on groundwater utilization in coastal areas.

Saltwater intrusion is one of the most widespread and important processes that degrade water quality to levels exceeding acceptable drinking and irrigation water standards, and endanger future water exploitation in coastal aquifers. Coupled with a continuing sea level rise due to global warming, coastal aquifers are even more under threat. This problem is intensified due to population growth, and the fact that about 70% of the world's population occupies the coastal plain zones [5,6]. The intensity of the problem depends on the amount of the abstraction, in relation to the natural groundwater recharge, as well as on the well field location and design, the geometry, and the hydrogeological parameters of the pumped aquifer.

In recent years, there is an increasing interest in evaluating the extent of seawater intrusion in response to overexploitation and sea level rise [7]. Seawater intrusion phenomena have been reported with different degree, in almost all coastal aquifers around the globe. In the United Sates, saltwater intrusion into coastal aquifers has been identified in the eastern Atlantic [8–14], and the southern [15] and western Pacific [16–19] coasts.

Reference [20] provides an overview of saltwater intrusion in the 17 coastal states of Mexico, which is one of the most important cases around the globe. Seawater intrusion induced by groundwater development is also known in South America [21] and Australia [22]. In Africa, several cases of seawater intrusion into coastal aquifers have been reported [23] with case studies in Morocco [24–26], Tunisia [27], Algeria [28] and Dar es Salaam [29,30]. In Europe, seawater intrusion has been documented within many of the coastal aquifers, particularly along the North Sea [31,32] and the Mediterranean Sea and its eastern part in the archipelagos in the Aegean Sea [33–35].

The arid and semi-arid areas are mostly chronically water-stressed. The problem of seawater intrusion is more severe in arid and semi-arid regions where the groundwater constitutes the main freshwater resource, which is mostly non-renewable. At present, developing countries of the Mediterranean Basin in North Africa and Middle East face environmental pressures induced by high population growth, rapid urbanization, and deficient water sector services reflecting on improper management of water resources [36,37]. The shortage of water in the Mediterranean region has been affected by the impact of climate change (increase of temperatures, variation of precipitations and high potential of evapotranspiration). Once again, the impacts have different effects in the Mediterranean region: the semi-arid and arid regions of the basin are exposed to desertification, increasing salinity of freshwater and exhaustion of water sources. Climatic change will also alter the marine environment, with an expected rise in sea level modifying several shores of the Mediterranean countries. To overcome the consequences of water scarcity and climate change, the aquifers and groundwater seem to be the solution. Many of the most important water projects in these regions focus on fossil water creating a sort of "pumping race" between the countries that share common aquifers, where overexploitation of groundwater in these regions is the major cause of seawater intrusion problems [34,35,38–46]. Therefore, the main challenges in coastal areas in the semi-arid region are water conservation, management and planning of the water resources. This is further complicated with several complexities of the geological formations. With the semi-arid conditions, complex geological settings and over-shooting stresses, the aquifer system becomes extremely fragile and sensitive [1]. Despite a good amount of research in this field, it is still needed to understand the behavior of such complex system precisely and apply the result in reasonably larger scales.

In the western Mediterranean, the situation of the groundwater in the Maghreb countries (Egypt, Libya, Tunisia, Algeria and Morocco) in North Africa has been marked by continuous decreases of water levels in coastal aquifers reaching alarming values. This decrease, caused by the synergistic effects of drought, flooding, changing land use, pollution from agriculture and industrialization, has intensified the problem of seawater intrusion [1]. The Korba coastal aquifer situated in Cap-Bon, Tunisia, has been experiencing seawater intrusion since 1970 and currently the salt load in this unconfined aquifer has peak concentrations of 5–10 g/L [27,47]. The Algerian coastal aquifers have also not escaped overexploitation with the Mitidja aquifer suffering from seawater intrusion [28], especially during the dry season. This aquifer system has a steady decline in water level in the order of 20–50 m per decade, which increases the rate of seawater intrusion on an annual basis [48]. The origin of water salinity on the Annaba coast (North East Algeria) is attributed to several factors such as the geological features of the region, the climate and the salt deposits. The salinity increases steadily when approaching the sea, and indicates the influence of marine water [49]. In Morocco, areas have been identified in which saltwater intrusion occurs (Temara-Rabat; Nador: [24]; Saidia: [25], however the aquifer system also contains marine deposits which contribute to the degradation of the groundwater quality. The rates of water abstraction in these areas have increased in the last 50 years, resulting in the lowering of the water table and eventually allowing seawater to intrude from coastal areas [50].

In the Nile delta, seawater intrusion has been observed 60 km inland as a result of excessive pumping [51]. An extensive saltwater body has developed from upper Egypt to eastern Libya in the past 50 years. The freshwater/saline water interface passes through the Qattara depression crossing the Libyan-Egyptian border and finally turning to the Southwest reaching the Tazerbo area, southeast Libya [1]. The development of the Siwa oasis from the deep Nubian Sandstone Aquifer is close to the freshwater/saline water interface, and could cause the saline water to intrude into the freshwater aquifer (Internationally Shared Transboundary Aquifer Resources Management [52]. The problem is further compounded since on the Libyan side large amounts of water are abstracted for urban development, causing saltwater intrusion along the Libyan coast. This overabstraction in combination with the sluggish flow of the Nubian Sandstone Aquifer causes the saline water body to encroach even further inland with considerable increases in salinity due to seawater intrusion and upconing of deep saline water [46,53].

Progressive seawater intrusion in the coastal aquifers of Libya has been experienced since 1930s because of its ever-increasing water demand from underground water resources. Since the 1960s, the risk of seawater intrusion is continuously threatening large coastal parts of the Jifarah Plain that forms one of the economically most significant areas in Libya, where TDS peaks up to 10 g/L are recorded [54]. Numerous irrigated regions are located near the coast, principally in the northern part of Jifarah Plain including Tripoli region, where extensive irrigated areas have been established in the late 1970s and have evolved into advanced agricultural production zones; these activities are primarily dependent on groundwater extraction.

In Tripoli, the seawater intrusion has steadily increased from 1960 to 2007, a period during which potable water was available from the aquifer. Since 1999, a loss of 60% in well production in the upper aquifer has been observed [55].

Because of the accelerated development of the coastal zone of Tripoli in the last decades, it is necessary to evaluate the saline water intrusion phenomena of the coastal aquifer to open different choices for the rational exploitation of the groundwater resources in this semi-arid zone avoiding the degradation of groundwater quality.

## 2. Introduction to the Study Area

Libya's coastal area is one of the important cases in the arid and semi-arid regions, it is a south Mediterranean country and has a shoreline extent of about 1750 km. Groundwater is the main source for potable, industrial and irrigation water because of its semi-desert climate. Inevitably, groundwater extraction has been in excess of replenishment because of the rapid increase in agricultural and economic activities in the last 50 years. This has resulted in water level decline and deterioration in quality, including invasion of seawater along the coastal regions.

This situation has led to two significant problems linked to human activity: (1) salinization due to the formation of large piezometric drawdown cones, which have induced seawater intrusion by reversing the hydraulic gradients into aquifers; and (2) direct input of nitrate mainly from fertilizers and sewage. Agriculture is based on intensive irrigation and fertilization to improve the soils.

The Jifarah Plain in the northwest of the country, located between the Mediterranean coast and the Jabel Naffusah Mountain in the south, contains more than 60% of the country's population and produces 50% of the total agricultural outputs. Tripoli area is a typical example showing the problems of coastal zones under high anthropogenic pressure in dryland regions. Tripoli city forms an almost rectangular area (763 km$^2$) between the Mediterranean Sea and the cities of Swani and Bin Gashir in the south (Figure 1). This area extends for about 20 km along Tripoli coastal area and about 22 km inland. Topography is rising towards the south and east, a general trend in overall Jifarah Plain, which is bounded to the south and east by Jebal Nafusseh Mountains. The shortage of good quality water from surface sources has made groundwater to be very important in the study area. The scarcity of water in Tripoli is becoming more pronounced due to the increase of the population coupled with improvement of the standard of living over the last few decades, where the area of the study accommodates dense population with more than 1.5 million of inhabitants mostly concentrated in the coast.

The principal aquifer used by the population in Tripoli is the Upper Miocene-Pliocene-Quaternary aquifer system, called "first aquifer" or "upper aquifer"; intercalated thin clayey sand and marl series are dividing the aquifer into a number of horizons, all are considered as one unconfined unit [56]. The Tripoli upper aquifer is affected by different sources of salinization, most serious is seawater intrusion [54]. The aim of this study was to discover what processes have been responsible for variations in the chemical composition of groundwater in the upper aquifer of Tripoli and to recognize the different sources of pollution, and their relation to the intense water withdrawal.

The climate in the study area is arid to semi-arid and typically Mediterranean, with irregular annual rainfall. The average annual rainfall and evapotranspiration rates are 350 mm/year and 1520 mm/year, respectively [54]. The estimation of groundwater exploitation (in the whole Jifarah Plain) from the main upper aquifer shows that the total amount of groundwater pumped in the Jifarah Plain for domestic, industrial and agricultural uses amounts to 1201.30 $Mm^3$/year [54]. For drinking water supply and domestic wells, the overall amount pumped is 6% of the total amount of groundwater extraction. The yield of irrigation wells was estimated to be 1123 $Mm^3$/year, which is equal to 93% of the total amount of groundwater extraction. The industrial sector pumps only 1% of the total groundwater exploitation in the plain. Since 1996, the Great Man-Made River Project is supplying the plain with an amount of 149 million $m^3$/year of water. This amount has been considered in the total abstraction estimation [54].

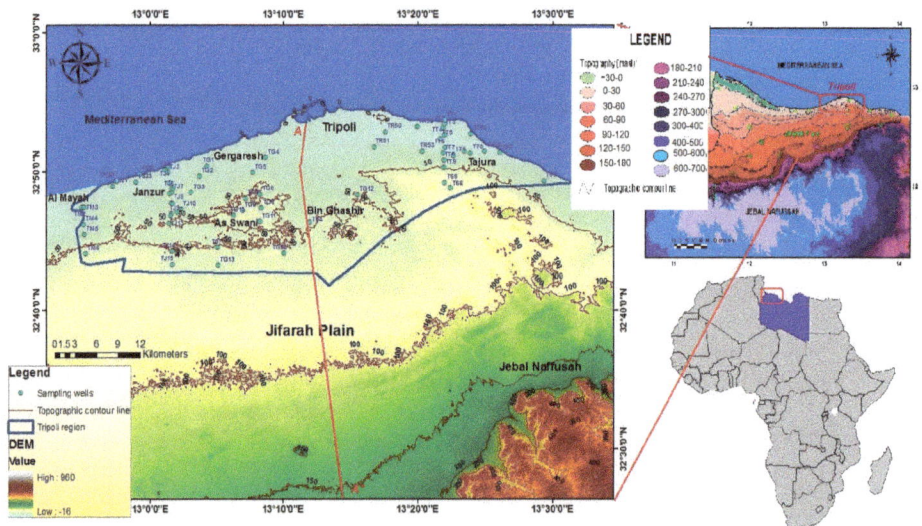

**Figure 1.** Location and topography of the study area in Jifarah Plain North West, Libya.

*Geological and Hydrogeological Setting*

The Jifarah Plain, including the study area, has been the subject of numerous geological studies [57–60].

Jifarah Plain is situated on the continental margin of Africa. Although the plain is thought to be underlain by Paleozoic rocks, the oldest encountered in boreholes are Triassic in age. These are continental, passing upwards into evaporites, a sequence thought to represent the progressive subsidence of the margin during major Mesozoic extension of the Tethyan Ocean. Continued subsidence through the Jurassic and Early Cretaceous led to the deposition of marine sequences.

Figure 2 shows the geological cross-section in Jifarah Plain crossing Tripoli region. The location of the cross-section is indicated in Figure 1. The sediments of the Jifarah plain have been deposited since early Mesozoic times in a near shore lagoonal environment. The lithology of the upper aquifer varies widely and includes detrital limestone, dolomite, gravel, marl, clay, silt, sand, sandstone, gypsum/anhydrite and calcarenite. Middle Miocene clay separates the upper aquifer system in the area from the middle aquifer. The depth to the bottom of the upper aquifer varies between 30 and 200 m and depths of the wells that are utilizing this aquifer are between 10 and 180 m. Most of the wells tapping this aquifer give productivity of about 20–80 $m^3$/h [56].

The geological deposits, that are playing a role in the hydrogeology of the area comprising the Upper-Miocene-Pliocene-Quaternary formations, are given in Table 1.

The Pleistocene formations include terraces, which consist of cemented gravel and conglomerate. Al Kums Formation consists of limestone and dolomite. Qasr Al Haj Formation is mainly alluvial fans and cones consisting of clastic materials derived from the scarp. Jeffara Formation consists mainly of silt and sand, occasionally with gravel caliche bands; it covers extensive parts of the Jifarah Plain. Gergaresh Formation, which is known as Gergaresh Sandstone of Tyrrhennian age, occasionally contains silt lenses, conglomerate and sandy limestone.

The Holocene deposits include recent wadi deposits; these deposits consist of loose gravels and loam. Beach sands are represented by a narrow strip at the coast and are made up of shell fragments with a small ratio of silica sands. Eolian deposits are represented by sand dunes and sheets covering large parts of the coastal strip (coastal dunes). These coastal dunes consist of shell fragments with small amounts of silica sands. It is worth mentioning that the eolian material composing coastal dunes contains a large amount of grains of gypsum. In some places, it is composed of nearly pure gypsum (98%) especially in the immediate vicinity of the sebkhas, with a silty gypsum filling [58]. Sebkha sediments are mainly gypsum deposits and are observed along the coastal area of the plain. They occupy the relatively low topographic areas and are separated from the sea by sea cliffs. Some of the sebkhas have occasional incursions of the sea and others may have subsurface connection with the seawater.

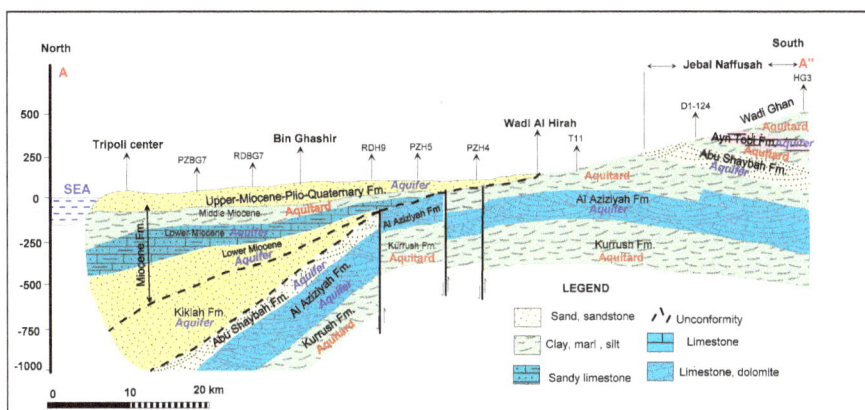

**Figure 2.** Geological cross-section in the coastal area of Jifarah Plain, including Tripoli region.

**Table 1.** Description of the Upper-Miocene-Pliocene-Quaternary deposits in the coastal area of Jifarah Plain around Tripoli.

| Period | Epoch | | Deposits and Formations | Typical Lithology | Thickness (m) |
|---|---|---|---|---|---|
| Quaternary | Holocene | Upper Miocene-Pliocene-Quaternary aquifer | Wadi deposits | Loose gravel, loam | 5–150 |
| | | | Sand beach | Shell and silica sand | |
| | | | Sand dunes and sand sheets | Shell fragments, silica sands and gypsum | |
| | | | Sebkha deposits | Gypsum | |
| | | | Fluvial-Eolian deposits | Silt, clay, marl and fine sand | |
| | | | Gergaresh Formation | Conglomerate, sandstone, silt, sandy limestone | |
| | Pleistocene | | Jeffara Formation | Silt, sand and gravel caliche bands | |
| | | | Qsar al Haj Formation | Alluvial fans and cones | |
| | | | Al Kums Formation | Limestone, dolomite | |
| | Miocene | | Volcanic rocks | Basalt and phonolite | 25–250 |
| | | Middle | Middle Miocene clay | | |

## 3. Methodology

### 3.1. Sampling and Analytical Methods

A regional hydrogeochemical survey and water level measurements were performed during the dry period from September to November of 2008. A total of 64 shallow and deep wells (mostly 10–180 m deep), located at different distances from the Mediterranean Sea, were selected for groundwater sampling and water level measurement (see Figure 1). The samples were collected during pumping and the water level measurements were performed beforehand in static condition.

The sampling points were chosen along vertical lines perpendicular to the coast (Figure 1), with lengths comprised between 1 and 20 km, in order to explore the aquifer from inland to the coast line. Water depth was measured from the ground surface using water level meter, and was converted into water level by subtracting from ground elevation. The collected water samples were preserved in polyethylene bottles after filtering with 0.45 μm cellulose membrane filters. Two samples were taken from each well, one for determining anions, the other for determining cations. Samples for cation analysis were acidified to lower the pH to around pH = 2 by adding a few drops of nitric acid. Parameters measured are physical properties such as: pH, temperature, water level and electrical conductivity. Cations ($Na^+$, $K^+$, $Mn^{2+}$, $Fe^{Total}$, $Ca^{2+}$, $Mg^{2+}$, $Zn^{2+}$, and $Si^{4+}$) were analyzed using Flame Atomic Absorption Spectrometry (Varian). Anions ($Cl^-$, $NO_3^-$, $NO_2^-$, $SO_4^{2-}$, and $PO_4^{3-}$) and $NH_4^+$ were analyzed using the Molecular Absorption Spectrophotometer (Shimadzu). $F^-$ was measured with ion selective electrode. Determination of carbonate ($CO_3^{2-}$) and bicarbonate ($HCO_3^-$) used the titration method with dilute HCl acid to pH 8.2 and 4.3, respectively. The above-mentioned analytical methods were used at the Laboratory of Applied Geology and Hydrogeology, Ghent University, and were provided in the Laboratory Manual and in Standard Methods for Examination of Water and Wastewater (American Public Health Association [61]. Careful quality controls were undertaken for all samples to obtain a reliable analytical dataset with an ionic balance error less than 5%.

### 3.2. Hydrochemical Evaluation Methods

The interpretation process is mainly based on the calculation of the ion deviations ($\Delta m_i$) from conservative freshwater/seawater mixing, the calculation of the saturation indices (SI), Stuyfzand classification system, graphical illustration methods including Piper diagram, calculation of ionic ratios, and elaboration of hydrochemical profile and maps showing the spatial and vertical distribution of water quality parameters in the study area.

### 3.2.1. Saturation Indices

The PHREEQC 2.16 program [62] was used to calculate saturation indices for calcite, dolomite, halite and gypsum based on the chemical analytical results and measured field temperatures for all samples.

### 3.2.2. Ion Deviation from Conservative Freshwater/Seawater Mixing

Calculation of the ionic deltas $\Delta m_i$ consists of a comparison of the actual concentration of each constituent with its theoretical concentration for a freshwater/seawater mix calculated from the $Cl^-$ concentration of the sample [63], because $Cl^-$ is the dominant ion in seawater and can be assumed to be conservative in many natural waters [64]. The ionic deltas quantify the extent of chemical reactions, affecting groundwater composition, next to mixing. The chemical reactions during fresh/seawater displacement can be deduced by calculating a composition based on the conservative mixing of seawater and freshwater, and comparing the conservative concentrations with those in the samples. The mass fraction of seawater ($f_{sea}$) in the groundwater can be obtained from chloride concentrations of seawater and freshwater as follows [64]:

$$f_{sea} = \frac{m_{Cl^-,sample} - m_{Cl^-,fresh}}{m_{Cl^-,sea} - m_{Cl^-,fresh}} \tag{1}$$

where

$m_{Cl^-,sample}$ = the concentration of $Cl^-$ in the sample expressed in mmol/L;
$m_{Cl^-,fresh}$ = the concentration of $Cl^-$ in the freshwater expressed in mmol/L; and
$m_{Cl^-,sea}$ = $Cl^-$ concentration in the seawater end member in mmol/L (for Mediterranean Seawater (possible end member), $m_{Cl^-,sea}$ = 645 mmol/L; Da'as and Walraevens, 2010).

Based on the conservative mixing of seawater and freshwater, the concentration of an ion $i$ ($m_i$) in the mixed waters was calculated using the mass fraction of seawater $f_{sea}$ as follows [64]:

$$m_i, mix = f_{sea} \cdot m_i, sea + (1 - f_{sea})m_i, freshs \tag{2}$$

where $m_i$ is concentration of an ion $i$ in mmol/L and subscripts mix, sea, and fresh indicate the conservative mixture, and end members seawater and freshwater, respectively. Any change in concentration $m_{i,reaction}$ ($\Delta m_i$) as a result of reactions (not mixing) then becomes:

$$\Delta m_i = m_i, reaction = m_i, sample - m_i, mix \tag{3}$$

where $m_{i,sample}$ is the actually observed concentration in the sample in mmol/L.

The deviation from the conservative fresh/seawater mixing is due to chemical reactions. A positive delta means that the ion has been added to the water e.g., due to desorption from the exchange complex. Adsorption will lead to negative delta.

Ions in infiltrating rainfall near the coast are often derived from sea spray, and only $Ca^{2+}$ and $HCO_3^-$ are added due to calcite dissolution [64]. All other ions are thus ascribed to seawater admixture. In this case, $m_{i,fresh}$ = 0 for all components except $Ca^{2+}$ and $HCO_3^-$.

The main end members used in the calculations for this study are the Mediterranean seawater and freshwater from the upper aquifer. For Mediterranean Seawater where $Cl^-$ = 645 mmol/L, the seawater fraction has been calculated as:

$$f_{sea} = \frac{m_{Cl^-,sample}}{645} \tag{4}$$

Table 2 shows the ion concentrations in rainwater, Mediterranean seawater end member and the freshwater end member in Tripoli (based on representative sample in Janzur TJ17).

Recharge water in the plain is the water flowing to the aquifer from the high topographic recharge area in the south (Jebal Naffusah). As no data were collected from the south border of the plain, the groundwater in the recharge area is expected to have the same composition as the freshwater samples collected from a nearby high topographic region, where the freshest water sample (i.e., sample TJ17) is considered as a reference sample to the composition of freshwater coming from the south. The recharge water in this sample has a high concentration of $Ca^{2+}$ and $HCO_3^-$ as a result of calcite dissolution. The analyzed recharge water in the plain is also showing considerable concentrations of $Na^+$, $Mg^{2+}$ and $SO_4^{2-}$ as a result of carbonate and evaporite dissolution in the unsaturated zone, and a great impact of concentration by evaporation, that is characteristic for the study area.

**Table 2.** Chemical composition of possible end members.

| Parameter (Unit mg/L) | Analyzed Rainwater in Jifarah Plain | Analyzed Recharge Water in Tripoli (TJ17) | Mediterranean Seawater [54] |
|---|---|---|---|
| pH | 7.64 | 7.97 | - |
| $Na^+$ | 31.50 | 43.75 | 12,700 |
| $K^+$ | 3.00 | 4.50 | 470 |
| $Ca^{2+}$ | 13.40 | 49.51 | 470 |
| $Mg^{2+}$ | 3.15 | 8.70 | 1490 |
| $Cl^-$ | 26.10 | 55.15 | 22,900 |
| $SO_4^{2-}$ | 10.33 | 36.62 | 3190 |
| $HCO_3^-$ | 102.48 | 174.46 | 173 |
| $NO_3^-$ | 1.12 | 29.10 | 0 |
| $NO_2^-$ | 0.16 | 0.001 | - |
| $PO_4^{3-}$ | 0.13 | 0.07 | - |
| Fe (Total) | 0.0 | 0.008 | - |
| $Mn^{2+}$ | 0.01 | 0.01 | - |
| $NH_4^+$ | 0.22 | 0.001 | - |
| TDS | 199.24 | 401.88 | 41,393 |

### 3.2.3. Stuyfzand Classification

The Stuyfzand classification [65–67] subdivides the most important chemical water characteristics at four levels: the main type, type, subtype, and class of a water sample (Tables 3–5). Each of the four levels of subdivision contributes to the total code (and name) of the water type.

The major type is determined based on the chloride content, according to Table 3. The type is determined based on an index for hardness (see Table 4), which can be expressed in French hardness degrees:

$$TH = 5 \times (Ca^{2+} + Mg^{2+}) \text{ in meq/L} \tag{5}$$

The classification into *subtypes* is determined based on the dominant cations and anions (Figure 3). First, the dominating hydrochemical family (and groups within families between brackets) is determined both for cations (Ca + Mg, (Na + K) + $NH_4$ or (Al + H) + (Fe + Mn)) and anions (Cl, $HCO_3$ + $CO_3$ or $SO_4$ + ($NO_3$ + $NO_2$)). The most important cation and anion (group: within a group: the dominant ion in that group) determine the name of the subtype. Finally, the *class* is determined based on the sum of $Na^+$, $K^+$ and $Mg^{2+}$ in meq/L, corrected for a sea salt contribution (Equation (6)). This indicates if cation exchange has taken place and also the nature of the exchange, by assuming that all $Cl^-$ originates from seawater, that fractionation of major constituents of the seawater upon spraying can be neglected and that $Cl^-$ behaves conservatively.

$$\{Na^+ + K^+ + Mg^{2+}\} \text{ corrected} = [Na^+ + K^+ + Mg^{2+}] \text{ measured} - 1.061Cl^- \tag{6}$$

where

    − = often pointing at a saltwater intrusion;

    + = often pointing at a freshwater encroachment; and

    0 = often pointing at an equilibrium.

Each of the subdivisions contributes to the total code (and name) of the water type (see Table 5); for example, B4–NaCl– reads as: "brackish extremely hard sodium chloride water, with a {$Na^+ + K^+ + Mg^{2+}$} deficit". This deficit is often due to cation exchange during saltwater intrusion (salinization). It is well known that the hydrogeochemical composition of coastal groundwater affected by seawater intrusion is mainly controlled by cation exchange reactions next to the simple mixing process [64]. These reactions can explain deviations of the concentrations of cations from conservative mixing of both waters.

**Table 3.** Water type classification [65].

| Main Type | Code | Cl (mg/L) |
|---|---|---|
| Fresh | F | $\leq 150$ |
| Fresh-brackish | Fb | 150–300 |
| Brackish | B | 300–1000 |
| Brackish-salt | Bs | 1000–10,000 |
| Salt | S | 10,000–20,000 |
| Hyperhaline | H | >20,000 |

Note: Division in main types based on chloride concentration.

**Table 4.** Subdivision of the main types based on hardness [65].

| Number | Name | Code | Total Hardness (mmol/L) | Natural Occurrence in Main Types |
|---|---|---|---|---|
| −1 | Very soft | * | 0–0.5 | F |
| 0 | Soft | 0 | 0.5–1 | F Fb B |
| 1 | Moderately hard | 1 | 1–2 | F Fb B Bs |
| 2 | Hard | 2 | 2–4 | F Fb B Bs |
| 3 | Very hard | 3 | 4–8 | F Fb B Bs |
| 4 | Extremely hard | 4 | 8–16 | Fb B Bs S |
| 5 | Extremely hard | 5 | 16–32 | Bs S H |
| 6 | Extremely hard | 6 | 32–64 | Bs S H |
| 7 | Extremely hard | 7 | 64–128 | S H |
| 8 | Extremely hard | 8 | 128–256 | H |
| 9 | Extremely hard | 9 | $\geq 256$ | H |

Note: * No code number.

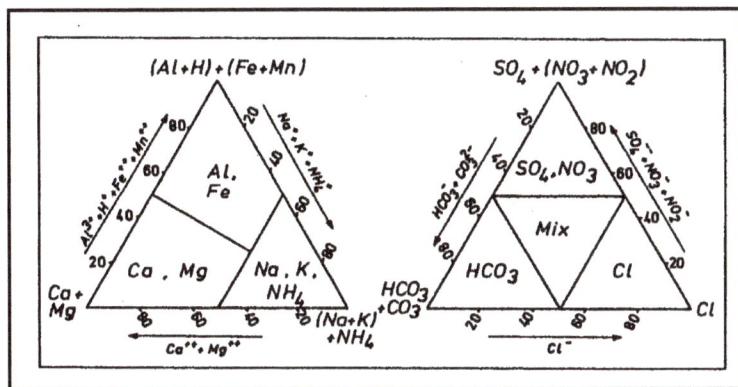

**Figure 3.** A ternary diagram showing the subdivision of types into subtypes [65].

**Table 5.** Subdivision of subtypes into classes according to {Na$^+$ + K$^+$ + Mg$^{2+}$} corrected for sea salt [65].

| Class | Code | Condition (meq/L) |
|---|---|---|
| {Na + K + Mg} deficit | - | {Na + K + Mg} corrected $< -\sqrt{0.5}$ Cl |
| {Na + K + Mg} equilibrium | 0 | $-\sqrt{0.5}$ Cl $\leq$ {Na + K + Mg} corrected $\leq +\sqrt{0.5}$ Cl |
| {Na + K + Mg} surplus | + | {Na + K + Mg} corrected $> \sqrt{0.5}$ Cl |

## 4. Results and Discussion

### 4.1. Water Level and Hydrodynamics

Figure 4 shows a piezometric map of Tripoli based on field measurements. In the coastal area, 64 shallow and deep wells located at different distances from the Mediterranean Sea were selected for water level measurement (see Figure 1). Water depth was measured from the ground surface using water level meter and was converted into water level by subtracting from ground elevation.

The overall direction of groundwater flow in Jifarah Plain in general, is from the south from Jebal Naffusah Mountains to the coast. The important storage withdrawal by overexploitation from the upper aquifer in Tripoli is causing continuous drawdown of the water level, reducing the outflow rate to the sea, and the progressive degradation of the chemical quality of water. Groundwater level is mostly low, especially near the coast, where zero and negative heads are recorded for the majority of wells. The piezometric level in depression cones at the location of the public water supply well field of As Swani (123 wells) has dropped from 25 to 33 m below sea level (Figure 4), which testifies the inversion of the hydraulic gradient and the intrusion of seawater.

From Figure 4, it can be concluded that the general groundwater flow, from south to north following the topography, has altered, where, generally along the coast, flow is toward the reduced heads in the stressed areas around the depression cones.

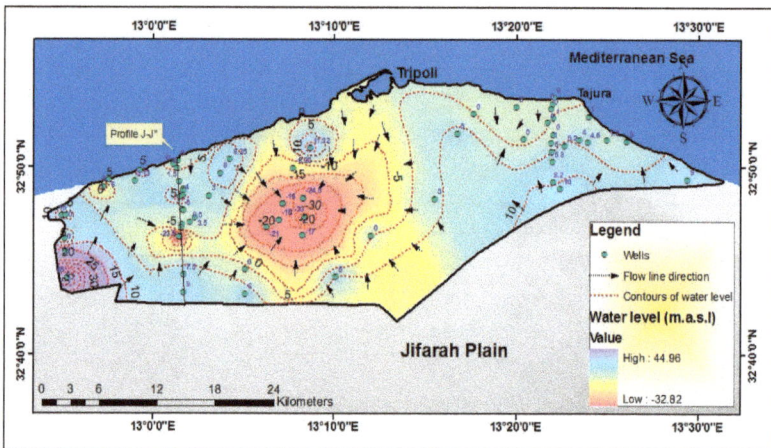

**Figure 4.** Tripoli piezometric map with flow vectors.

To deal with the shortage of water in most coastal cities including Tripoli region, the Libyan government established the Great Man-Made River Project (GMMR) to transport millions of cubic meters of water a day from desert well fields to the coastal cities, where over 80% of the population lives. Since 1996, The GMMR is supplying Tripoli city with an amount of 149 million m$^3$/year. The total planned supply by the project is 900 million m$^3$/year for the whole Jifarrah Plain. The implementation of the project was interrupted since 2011, due to political situation and Tripoli is the only supplied part of Jifarrah coast.

Since the start of the supply in 1996, the wellfield of As Swani (123 public water supply wells) was stopped. The pumping from As Swani wellfield is the main cause of the depression cone in Tripoli; the depression cone location at the center of As Swani wellfield is the most affected part of the region. Now the GMMR is the main supplier for domestic use in the city and being used also in many farms. The drawdown value around the depression cone is being reduced from 80 m below sea level in 1996 to 33 m in 2008.

*4.2. Major Hydrochemical Parameters*

Major anions and cations were analyzed and pH, Eh, electrical conductivity (EC) as well as temperature were assessed on all samples. The results show that: temperature ranges 18–25 °C, pH range is 7.17–9.94, Eh range is 139–240 mV, EC range is 369–10,600 μS/cm (25 °C), TDS range is 340–6529 mg/L and chloride concentration ranges 39–3155 mg/L. Table 6 shows analytical results for physico-chemical parameters of groundwater for selected representative samples in Tripoli. The high concentration of major ions such as Cl, Na, Ca, K, Mg, and a high EC indicate the presence of seawater in an aquifer [68,69].

Out of 64 samples analyzed, 38% have $NO_3^-$ higher than the highest desirable level of 45 mg/L [70]. For sulfate 40% have $SO_4^{2-}$ higher than the highest desirable level of 200 mg/L according to WHO (2008), with a maximum of 835 mg/L recorded southwards. Out of 64 analyzed samples, 58% exceed the recommended $Cl^-$ value for standard drinking water (250 mg/L) and 26% have $Cl^-$ greater than the highest admissible level of 600 mg/L [70].

Levels of $Cl^-$ and EC are the simplest indicators of seawater intrusion or salinization [71,72]. EC is positively correlated with the concentration of ions, mainly $Cl^-$ concentration. Figure 5 shows three zones on a plot of $Cl^-$ vs. EC: freshwater zone, mixing zone and strong mixing (intrusion). It shows that groundwater samples with $Cl^-$ exceeding 200 mg/L and EC exceeding ~1000 μS/cm are most likely influenced by seawater intrusion. Groundwater samples that are characterized by EC between 1000 and 5000 μS/cm represent a mixing between freshwater and saltwater. Samples with EC of more than 10,000 μS/cm represent strong seawater influence.

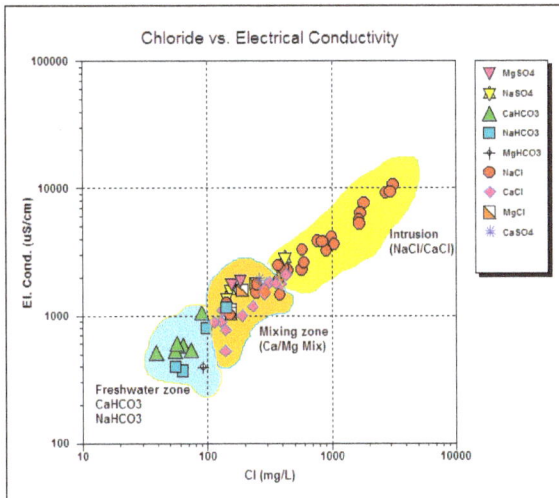

**Figure 5.** A plot of chloride vs. electrical conductivity showing fresh groundwater conditions, saltwater intrusion, and mixing between the two end members.

**Table 6.** Analytical results for physico-chemical parameters of selected groundwater samples in Tripoli.

| ID | T (°C) | Water Level (m a.s.l) | pH | Eh (mV) | EC (μ/cm 25°) | Ca²⁺ mg/L | Mg²⁺ mg/L | Na⁺ mg/L | K⁺ mg/L | Mn²⁺ mg/L | Fe²⁺/³⁺ mg/L | NH₄⁺ mg/L | NO₃⁻ mg/L | NO₂⁻ mg/L | CO₃²⁻ mg/L | HCO₃⁻ mg/L | Cl⁻ mg/L | SO₄²⁻ mg/L |
|---|---|---|---|---|---|---|---|---|---|---|---|---|---|---|---|---|---|---|
| TM5 | 22 | 19.11 | 7.30 | 234 | 2840 | 200 | 104 | 334 | 14 | 0.01 | 0.40 | 0.07 | 21 | 0.01 | 0 | 273 | 419 | 837 |
| TM6 | 22 | 45.20 | 7.45 | 234 | 2500 | 110 | 72 | 283 | 6 | 0.01 | 0.41 | 0.19 | 118 | 0.01 | 0 | 219 | 371 | 382 |
| TS21 | 21 | −8.01 | 7.97 | 133 | 1160 | 82 | 42 | 99 | 8 | 0 | 0 | 0 | 11 | 0 | 6 | 287 | 142 | 91 |
| TJ1 | 24 | 0 | 8.22 | 160 | 1826 | 172 | 29 | 186 | 16 | 0.01 | 0.01 | 0 | 16 | 0 | 24 | 269 | 316 | 221 |
| TJ2 | 24 | 0 | 7.82 | 142 | 1940 | 93 | 84 | 160 | 6 | 0 | 0 | 0.02 | 69 | 0.11 | 0 | 323 | 376 | 123 |
| TJ3 | 23 | 0 | 7.40 | 152 | 2380 | 140 | 93 | 184 | 14 | 0 | 0.01 | 0 | 83 | 0 | 0 | 317 | 410 | 268 |
| TJ4 | 24 | 1.50 | 8.17 | 188 | 891 | 78 | 29 | 67 | 4 | 0 | 0 | 0 | 16 | 0 | 1 | 189 | 128 | 62 |
| TJ5 | 24 | 2.25 | 8.16 | 143 | 1100 | 82 | 44 | 81 | 8 | 0 | 0 | 0 | 20 | 0 | 12 | 208 | 131 | 144 |
| TJ6 | 22 | 1.50 | 7.59 | 161 | 524 | 71 | 40 | 76 | 5 | 0 | 0 | 0.03 | 73 | 0.11 | 0 | 236 | 138 | 87 |
| TJ7 | 23 | 4.00 | 8.2 | 139 | 510 | 40 | 19 | 41 | 4 | 0 | 0 | 0 | 5 | 0 | 18 | 140 | 39 | 53 |
| TJ8 | 23 | 3.80 | 8.19 | 156 | 878 | 62 | 36 | 67 | 4 | 0 | 0 | 0 | 18 | 0 | 12 | 165 | 113 | 96 |
| TJ10 | 24 | 3.50 | 9.94 | 142 | 1020 | 66 | 41 | 79 | 5 | 0 | 0.08 | 0 | 53 | 0.14 | 0 | 198 | 151 | 115 |
| TJ11 | 22 | 0 | 7.54 | 169 | 1530 | 116 | 65 | 90 | 6 | 0 | 0.12 | 0 | 79 | 0.10 | 0 | 220 | 290 | 148 |
| TJ12 | 22 | 3.00 | 7.85 | 140 | 1592 | 72 | 64 | 111 | 6 | 0.01 | 0.04 | 0.10 | 46 | 0.11 | 0 | 190 | 189 | 161 |
| TJ13 | 24 | −23.90 | 7.77 | 150 | 1764 | 147 | 77 | 96 | 6 | 0.01 | 0.07 | 0.10 | 52 | 0.12 | 0 | 244 | 391 | 129 |
| TJ14 | 24 | 7.50 | 8.20 | 198 | 1928 | 188 | 53 | 175 | 8 | 0 | 0 | 0 | 11 | 0 | 12 | 281 | 259 | 413 |
| TJ15 | 23 | 9.00 | 7.30 | 175 | 1868 | 123 | 78 | 152 | 6 | 0 | 0.03 | 0.04 | 55 | 0.10 | 0 | 287 | 182 | 378 |
| TJ17 | 23 | 23.00 | 7.97 | 213 | 532 | 50 | 9 | 44 | 5 | 0.01 | 0 | 0 | 29 | 0 | 0 | 175 | 55 | 37 |
| TG1 | 19 | 26.35 | 7.75 | 149 | 2270 | 117 | 91 | 206 | 4 | 0.01 | 0.09 | 0.04 | 61 | 0.37 | 0 | 284 | 441 | 117 |
| TG6 | 22 | −24.50 | 7.40 | 151 | 7660 | 192 | 172 | 1023 | 22 | 0.01 | 0.16 | 0.04 | 77 | 0.12 | 0 | 232 | 1831 | 297 |
| TR52 | 22 | 0 | 7.95 | 166 | 6380 | 132 | 137 | 1049 | 31 | 0 | 0 | 0 | 38 | 0 | 39 | 384 | 1709 | 274 |
| TT2 | 22 | 0 | 7.73 | 148 | 9150 | 572 | 209 | 1129 | 70 | 0 | 0 | 0 | 0.80 | 0 | 6 | 421 | 2726 | 512 |
| TT3 | 22 | 0 | 8.22 | 142 | 3850 | 262 | 100 | 377 | 67 | 0 | 0 | 0 | 1 | 0 | 24 | 433 | 766 | 418 |
| TT6 | 21 | 0 | 7.73 | 162 | 9310 | 572 | 203 | 1178 | 20 | 0 | 0 | 0 | 23 | 0 | 3 | 226 | 3003 | 293 |
| TT7 | 21 | 1.00 | 7.97 | 168 | 3810 | 204 | 108 | 480 | 24 | 0 | 0 | 0 | 6 | 0 | 12 | 256 | 840 | 567 |
| TT9 | 24 | 5.80 | 8.30 | 183 | 600 | 50 | 24 | 48 | 8 | 0 | 0 | 0 | 7 | 0 | 9 | 189 | 57 | 67 |
| T63 | 23 | 9.00 | 8.03 | 191 | 582 | 52 | 18 | 53 | 4 | 0 | 0 | 0 | 6 | 0 | 77 | 64 | 64 | 53 |

The spatial distribution of EC and Cl⁻ from analyzed groundwater samples across Tripoli city is presented in Figure 6a,b. In general, the EC, which is tightly linked to TDS, is a measure of salinity, and therefore is generally closely related to the Cl⁻ content. Both EC and Cl⁻ show the same general decrease from the Mediterranean shoreline towards the south. Frequent local increases in both variables are observed at the depression cones as a result of the high pumping rate.

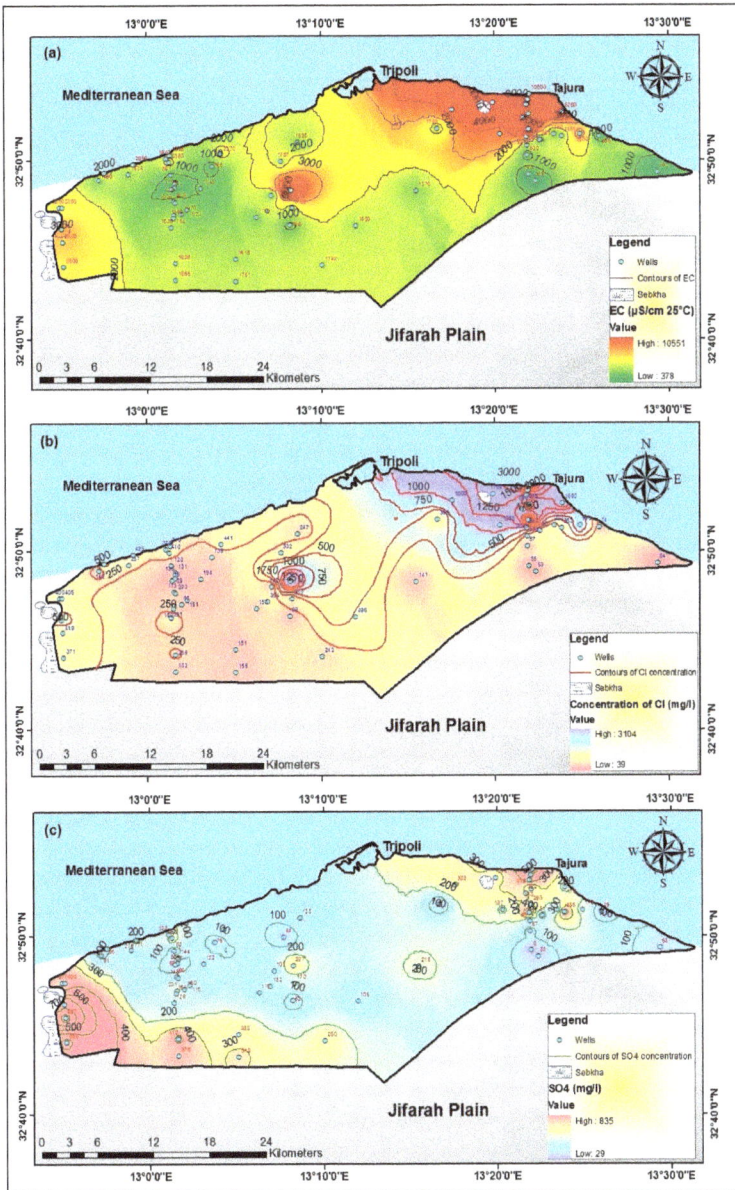

**Figure 6.** (a) Electrical conductivity map of Tripoli; (b) map with the spatial distribution of concentrations of Cl⁻; and (c) spatial distribution of $SO_4^{2-}$ in the upper aquifer of Tripoli.

The high Cl⁻ concentration is due to mixing with seawater. These high concentrations of chloride occur in most wells within a few kilometers of the coast, and are related to active seawater intrusion. However, high Cl⁻ is also found far inland at and nearby the depression cones, whereby deeper saline groundwater is affected by upconing due to groundwater exploitation. The concentration of Cl⁻ decreases gradually towards the south. However, in many farther inland wells, it is still the dominant anion. The higher concentration of Cl⁻ than the 250 mg/L value for standard drinking water at the south of the region can be linked to the synsedimentary marine influence of the groundwater.

Sulfate concentration in Tripoli ranges from 29 to 835 mg/L. Figure 6c shows the spatial distribution of $SO_4^{2-}$ concentration in the study area. The main source for increasing $SO_4^{2-}$ is mixing with seawater, which can add significant amounts of sulfate to freshwaters. High $SO_4^{2-}$ is mostly linked to the high Cl⁻ concentration in the upper aquifer, both in the seawater intrusion zones and in the depression cones, where deep saline water upconing occurs.

Besides, more than 500 mg/L $SO_4^{2-}$ is observed towards the west of region, with much higher $SO_4^{2-}/Cl^-$ compared to seawater. The main source of $SO_4^{2-}$ in this area is the dissolution of gypsum/anhydrite from the superficial sebkha deposits in those areas, as these wells are located near the vicinity of sebkhas. In this zone, lower Cl⁻ is recorded, excluding seawater intrusion as the source.

### 4.3. Water Types and Piper Diagram

Classification of hydrochemical facies for groundwaters according to the Piper diagram is represented in Figure 7.

In the Piper diagram, almost all water samples are plotted above the general seawater–freshwater mixing line [64], comprising freshwater sample TJ17 and Mediterranean Seawater. Although various hydrochemical facies were observed (NaCl, CaCl, MgCl, CaMgHCO₃, NaHCO₃, NaSO₄ MgSO₄ and CaSO₄), CaCl and NaCl types are dominant. Large proportion of the groundwater shows NaCl type, which generally indicates a strong seawater influence [50] or upconing of deep saltwater, while CaCl water type indicates salinization and cation exchange reaction [73]. The region of the CaCl type water may be a leading edge of the seawater plume [64,74,75]. Furthermore, sources of CaSO₄ water type are the dissolution of the scattered sebkha deposits.

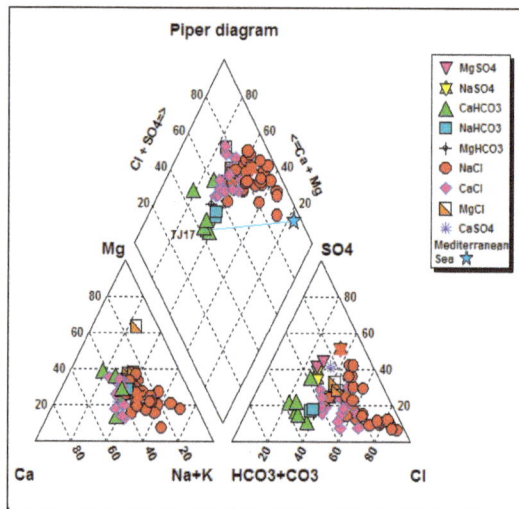

**Figure 7.** Water types according to Piper diagram.

*4.4. Hydrochemical Profile and Water Classification According to Stuyfzand*

Salinization is induced as the new saline end member is introduced into the freshwater aquifer. The main chemical reaction is cation exchange, resulting in deficit of $Na^+$ and surplus of $Ca^{2+}$:

$$Na^+ + 0.5Ca\text{-}X_2 \rightarrow Na\text{-}X + 0.5Ca^{2+}$$

where X represents the natural exchanger in the reaction. During cation exchange, the dominant $Na^+$ ions are adsorbed and $Ca^{2+}$ ions released, so that the resulting water moves from NaCl to CaCl water type, which is typical for salinization [5]. The salinization process can be schematized as follows [75] (water types according to classification of Stuyfzand, [65]):

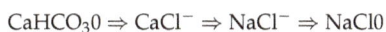

$$CaHCO_30 \Rightarrow CaCl^- \Rightarrow NaCl^- \Rightarrow NaCl0$$

$$F \text{ (fresh)} \Rightarrow Fb \text{ (fresh-brackish)} \Rightarrow B \text{ (brackish)} \Rightarrow Bs \text{ (brackish-saline)} \Rightarrow S \text{ (saline)}$$

The chloride ion concentration is taken as a reference parameter [64]. Therefore, as saltwater intrudes into coastal freshwater aquifers, the Na/Cl ratio decreases and the Ca/Cl ratio increases.

Upon the inflow of freshwater, a reverse process takes place:

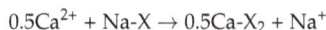

$$0.5Ca^{2+} + Na\text{-}X \rightarrow 0.5Ca\text{-}X_2 + Na^+$$

Flushing of the saline aquifer by freshwater will thus result in uptake of $Ca^{2+}$ by the exchanger with concomitant release of $Na^+$. This is reflected in the increase of the Na/Cl ratio, and formation of the $NaHCO_3$ water type, which is typical for freshening. The anion $HCO_3^-$ is not affected because natural sediments behave as cation exchanger at the usual near-natural pH of groundwater [64]. The freshening process can be schematized as follows [73]:

$$NaCl0 \Rightarrow NaCl^+ \Rightarrow NaHCO^{3+} \Rightarrow MgHCO^{3+} \Rightarrow CaHCO^{3+} \Rightarrow CaHCO^30$$

$$S \Rightarrow Bs \Rightarrow B \Rightarrow Fb \Rightarrow F$$

The major hydrogeochemical processes occurring in the upper aquifer are: mixing with seawater end member, cation exchange during salinization, dissolution of gypsum from superficial sebkha sediments, carbonate dissolution and agricultural pollution.

The hydrogeochemical profile in Tripoli (Janzur) (Figure 8) is selected as an example showing the distribution of water types along the flow path. Janzur profile is about 14 km long perpendicular to the sea (see Figure 2). Seventeen wells were visited in this profile region. Their total depth is shallow (between 10 and 50 m) close to the coast, and reaches 120 m southward. They are pumping for irrigation in the private farms and for domestic use. The groundwater table in the profile is between 0 and 4 m a.s.l in the north, while it is down to −24 m a.s.l in the south at the depression cone. Groundwater is flowing from the south and north to the locally reduced heads, located at about 10 km from the coast.

The spatial distribution of water sub-types according to Stuyfzand in Tripoli is presented in Figure 9 together with the pie plot of TDS distribution for selected representative samples in Tripoli region. The water classification scheme of Stuyfzand [65] has the advantage that in brackish or saline groundwater, one can still identify many different water types even though the major anions and cations are the same and this may help to recognize processes such as upconing of more saline water in the aquifer [67]. Figures 8 and 9 show water type is $CaSO_4$ in the west, $CaHCO_3$, $NaHCO_3$, CaMix and MgMix towards the south and NaCl, MgCl and CaCl in the north and at the depression cones. Close to the shoreline and mainly in the east in Tajura the water is NaCl type, due to the strong effect of seawater intrusion. CaCl results from cation exchange, due to mixing with seawater. Towards the south, CaMgMix($ClHCO_3$) evolving further inland to CaMgMix($HCO_3Cl$), indicates the location of the transition zone, where the groundwater changes from CaMgMix enriched with $Cl^-$ ion to CaMix with HCO$_3^-$ as the dominant anion. The CaSO$_4$ water type observed in western Tripoli, up to about 14 km inland, shows the dissolution of the evaporitic rocks from the sebkha deposits.

The Mg$^{2+}$ content found in several wells, is mainly resulting from the freshwater end member coming from the recharge area, where Mg$^{2+}$-containing carbonate is dissolved [54]. Thus, in this case, thee positive cation exchange code in the classification does not indicate freshening, as Mg$^{2+}$ is not supplied by the marine end member [69]. At the depression cones and downstream, cation exchange equilibrium (cation exchange code "0") exists for several wells, which in this case indicates the onset of the salinization process (the positive value of (Na$^+$ + K$^+$ + Mg$^{2+}$)$_{corrected}$ is decreasing as the marine cations are adsorbed during salinization).

The increase of salinity in Tripoli is accompanied by an increase in NO$_3^-$ concentrations. Figure 8 also shows the spatial distribution of NO$_3^-$ along Janzur profile. The average nitrate concentration of groundwater in the aquifer is about 38 mg/L, but contents as high as about 118 mg/L occur upstream in the south of the region. Irrigation with nitrogen fertilizers and domestic sewage and movement of contaminants in areas of high hydraulic gradients within the drawdown cones probably are responsible for localized peaks of the nitrate concentration for many wells in the region.

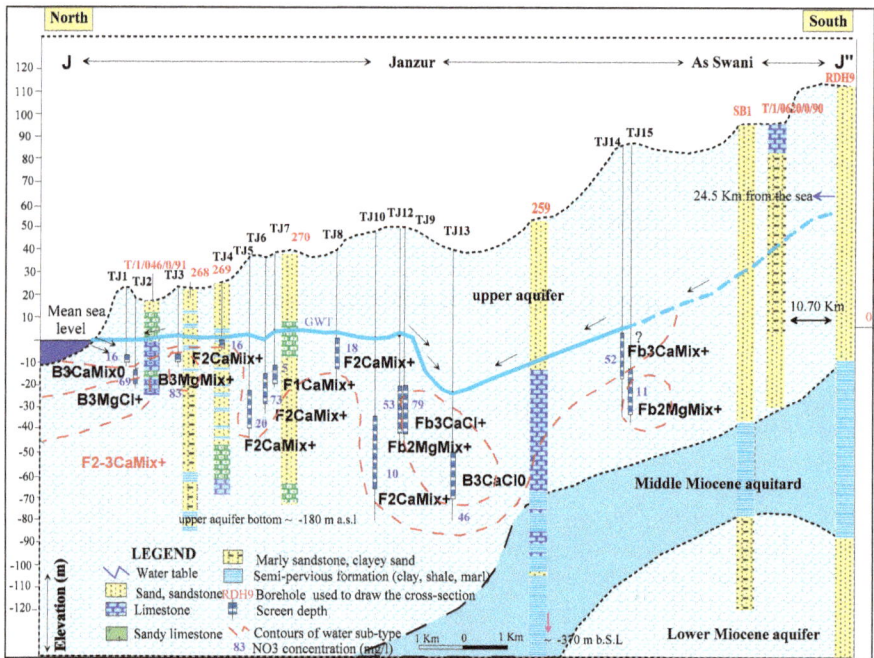

**Figure 8.** Hydrochemical profile in Tripoli (location of cross-section is indicated in Figure 2).

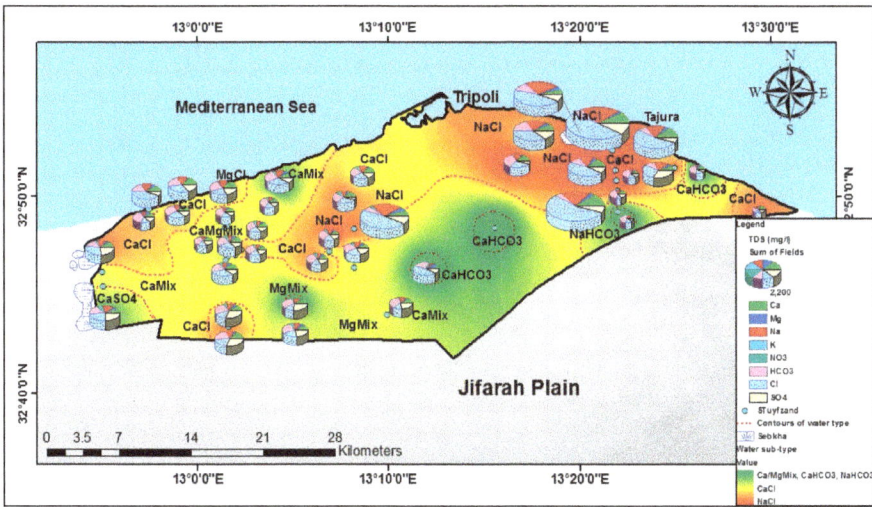

**Figure 9.** Distribution of water types according to Stuyfzand with a plot of TDS concentration for selected representative samples in Tripoli.

## 4.5. Ionic Ratio

Conservative seawater–freshwater mixing is expected to show a linear increase in $Na^+$ and $Cl^-$ [76], which is reflected by the high correlation coefficient ($r$ = 0.94) between both variables (Figure 10a). Effects of seawater encroachment have been evaluated by studying the Na/Cl ionic ratio. Lower ratios of Na/Cl than seawater values (0.88) indicate seawater encroachment. Figure 10b shows molar ratios of $Cl^-$ versus Na/Cl concentrations. The Na/Cl ratios for the analyzed samples range from 0.23 to 1.68. Most of groundwater samples were less than or slightly higher than the Mediterranean seawater ratio (0.88). The lowered values with respect to the Mediterranean seawater ratio are resulting from cation exchange occurring when seawater intrudes freshwater aquifers, resulting in the deficit of $Na^+$ and surplus of $Ca^{2+}$. High ratio for several samples towards the recharge area (e.g., TJ17) indicates flushing the aquifer by freshwater from the south.

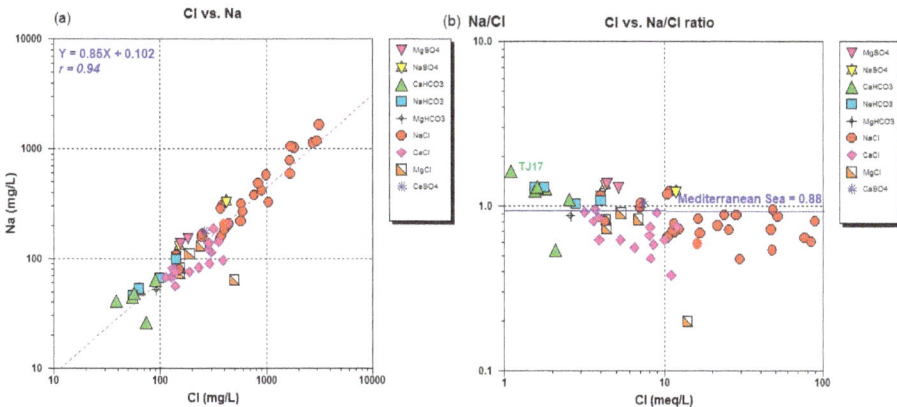

**Figure 10.** (a) Plot of $Cl^-$ vs. $Na^+$; and (b) molar ratio of $Cl^-$ vs. Na/Cl concentrations.

The ratio of $SO_4/Cl$ (meq/L/meq/L) for the Mediterranean seawater and the fresh recharge water from the study area are 0.54 and 0.25, respectively. The ratio of $SO_4/Cl$ for the analyzed samples ranges from 0.04 to 0.77, which indicates mixing between seawater and freshwater. Significantly higher values than Mediterranean seawater ratio (0.54) indicate dissolution of gypsum from Sebkha deposits.

The Na/K ratio is ranging in the area between 9.64 and 102.66, the largest values are observed in the area affected by the intrusion of seawater.

### 4.6. Saturation Indices

The saturation indices (SI) for calcite, dolomite, halite, aragonite, gypsum and anhydrite were calculated to verify precipitation and dissolution of these minerals. The selected minerals were based on the major ions in groundwater from the study area. Figure 11 is a synthetic diagram showing SI values for calcite, dolomite, gypsum, anhydrite, halite and aragonite. The sample numbers are sorted according to their location from west to east and from north to south with each profile, but they are not plotted at spatial distances.

Out of 64 groundwater samples, 80% of groundwaters seem to be supersaturated (SI > 0) with respect to calcite ($CaCO_3$), whereas 10% are undersaturated with respect to calcite (SI < 0) and 10% are at equilibrium (SI = 0). Dolomite ($MgCa(CO_3)_2$) seems to be oversaturated in 85% of groundwater samples analyzed, 5% are undersaturated and 10% are at equilibrium. Ninety-eight percent of groundwater samples in the study area are undersaturated with respect to gypsum ($CaSO_4 \cdot 2H_2O$) and anhydrite ($CaSO_4$).

In general, most of the analyzed samples have saturation indices close to saturation with respect to calcite (SI mostly 0–1) and dolomite (SI mostly 0–3). This slight supersaturation with respect to calcite and dolomite supersaturation rather points to groundwater in equilibrium with those minerals. During sampling, most often dissolved $CO_2$ gas escapes, slightly raising pH and thus shifting carbonate equilibrium (more $CO_3^{2-}$), such that SI > 0 is obtained, whereas in water in the aquifer SI with respect to calcite is close to zero. Thus, the water in the aquifer is not really oversaturated.

The majority of samples in the study area are undersaturated with respect to gypsum and anhydrite. Gypsum comes close to saturation (SI > −0.50) in several wells. The dissolution of gypsum from the superficial sebkha deposits for many wells in the coastal area raises the SI.

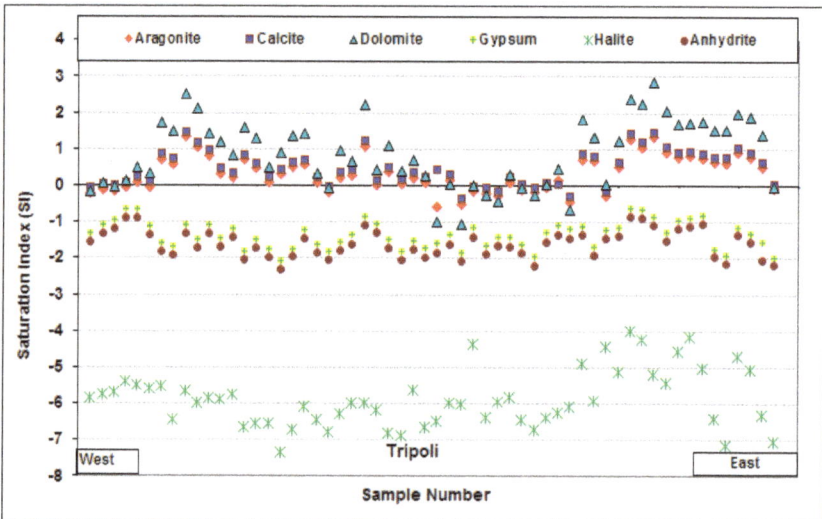

**Figure 11.** Calculated saturation indices of groundwater samples with respect to selected minerals.

## 4.7. Deviation from Conservative Mixture of End Member Fraction

Figure 12 shows the ionic deltas calculated for $Na^+$, $Ca^{2+}$, $Mg^{2+}$, $K^+$, $HCO_3^-$ and $SO_4^{2-}$ for all analyzed samples. The first thing to note is that the process of cation exchange due to salinization is very evident. For example, in Figure 12, $m_{Na^+,reaction}$ ($\Delta Na^+$) is plotted in the secondary axis; the $\Delta Na^+$ is usually positive for freshwater, but a large number of samples have negative values particularly in the highly saline water, down to $-23$ mmol/L. The most logical explanation for this deficit of $Na^+$ is that a reverse cation exchange reaction is taking place during the salinization process, which releases $Ca^{2+}$ to the solution and captures $Na^+$. The reverse relationship between the two ions ($Na^+$ and $Ca^{2+}$) is noticed particularly in the highly saline groundwater, where samples with large negative values of $\Delta Na^+$ generally show strong positive $\Delta Ca^{2+}$. Furthermore, also potassium shows negative (or very low positive) deltas characteristic for marine cations as a result of the salinization process.

$\Delta Mg^{2+}$ is mostly positive, due to more $Mg^{2+}$ added by dissolution of $Mg^{2+}$-rich carbonate than adsorbed at the clay exchange complex during salinization. Only very few samples show a deficit of $Mg^{2+}$. Figure 12 also shows that the ionic delta of $\Delta HCO_3^-$ is positive for most water samples. This is due to the dissolution of carbonate minerals in the aquifer deposits. In general, most samples show positive $\Delta SO_4^{2-}$. The gypsum dissolution from sebkha deposits increases $\Delta SO_4^{2-}$ to high positive values for several wells.

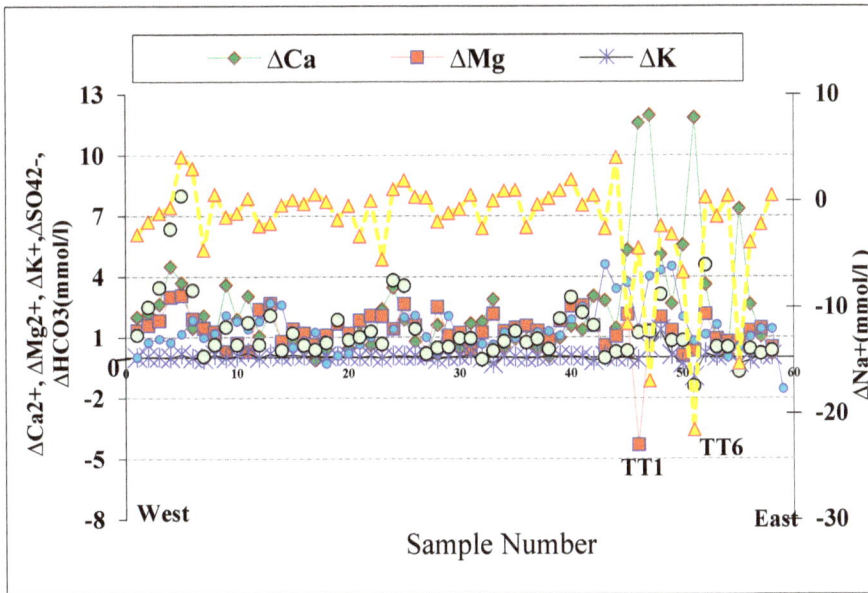

**Figure 12.** Diagram of ionic delta for all analyzed samples.

## 5. Conclusions

Hydrochemistry of the coastal aquifers of arid and semi-arid regions is very complex. Extensive groundwater extraction in Tripoli region, mainly for agricultural development, has caused substantial seawater encroachment and upconing of the deep saline water into Tripoli shallow aquifer, with $Cl^-$, $SO_4^{2-}$ and $NO_3^-$ as the major pollutants. In this study, this potential problem is investigated. The dominant water types in the study area are NaCl, CaCl and CaMgMix(ClHCO$_3$) except for several wells towards the recharge area, where CaHCO$_3$-type prevails, and wells located near the superficial sebkha deposits, where the CaSO$_4$ water type evolves.

Seawater intrusion is accompanied by chemical reactions, which modify the hydrochemistry of the coastal aquifer. The most remarkable reaction is that of the inverse cation exchange, characteristic of the changes of the theoretical mixture of seawater–freshwater, which is carried out between clays and the aquifer water. This exchange consists in the release of $Ca^{2+}$ and the adsorption of $Na^+$.

Great part of the observed high concentration of sulfate in Tripoli is coming from the effect of seawater intrusion. Furthermore, the scattered sebkha deposits in the north, containing large amounts of gypsum, produce high $SO_4^{2-}$ waters.

Another serious problem in the study area is the increased level of nitrate concentrations. It has been found that in large number of samples, nitrate contents range between 45 and 160 mg/L. The increased content originates from leaching of nitrates from the applied nitrogen fertilizers and from sewage. It is recommended that risk assessment of nitrate pollution is useful for a better management of groundwater resources, aiming at preventing soil salinization and minimizing nitrate pollution in groundwater.

The hydrochemical interpretation also indicates that the dissolution of calcite, dolomite and/or $Mg^{2+}$ bearing calcite is an important process in most of the groundwaters. The saturation index shows mostly a slight tendency to precipitation of calcite and dolomite in the aquifer system, but this can be ascribed to lowering of $CO_2$ pressure at sampling, while in the aquifer, there is equilibrium with respect to these minerals.

Although the Great Man Made River Project is supplying Tripoli with an amount of 149 million $m^3$/year of water since 1996, used mainly for domestic purposes, results show high degradation level of groundwater quality and most of water samples do not compare favorably with WHO standards [70]; many samples exceed the maximum admissible concentrations, highlighting the degradation of groundwater quality. The recovery of groundwater quality is usually a very slow process as seawater intrusion is the result of a long-term negative mass balance in the aquifer. A balance between pumping demand and quality requirements is necessary. This balance is hard to maintain when the final goal is to reverse the qualitative status of the already contaminated aquifer. To protect the groundwater resource in the long-term, on which the future Tripoli residents depend, appropriate management against overexploitation from agricultural activity to control salinity is compelling, and especially urgent in the coastal fringe, where seawater intrusion is threatening. Artificial recharge of coastal aquifers, which are especially overexploited, may offer an efficient means of combating seawater intrusion and thus of preventing an inevitable degradation of the water quality which might prove irreversible.

**Acknowledgments:** This study was supported by the Libyan government through the Libyan Embassy in Brussels. Great thanks to the well owners and all who supported in the field campaigns.

**Author Contributions:** Alfarrah Nawal conceived and designed the research. Walraevens Kristine supervised the study.

**Conflicts of Interest:** The authors declare no conflict of interest.

## References

1. Gaaloul, N.; Pliakas, F.; Kallioras, A.; Schuth, C.; Marinos, P. Simulation of Seawater Intrusion in Coastal Aquifers: Forty Five'Years exploitation in an Eastern Coast Aquifer in NE Tunisia. *Open Hydrol. J.* **2012**, *6*, 31–44. [CrossRef]
2. Masciopinto, C. Simulation of coastal groundwater remediation: The case of Nardò fractured aquifer in Southern Italy. *Environ. Model. Softw.* **2006**, *21*, 85–97. [CrossRef]
3. Mjemah, I.C.; Van Camp, M.; Walraevens, K. Groundwater exploitation and hydraulic parameter estimation for a Quaternary aquifer in Dar-es-Salaam, Tanzania. *J. Afr. Earth Sci.* **2009**, *55*, 134–146. [CrossRef]
4. Van Camp, M.; Mtoni, Y.E.; Mjemah, I.C.; Bakundukize, C.; Walraevens, K. Investigating seawater intrusion due to groundwater pumping with schematic model simulations: The example of the Dar Es Salaam coastal aquifer in Tanzania. *J. Afr. Earth Sci.* **2014**, *96*, 71–78. [CrossRef]

5.  Jones, B.F.; Vengosh, A.; Rosenthal, E.; Yechieli, Y. Geochemical investigation of groundwater quality. In *Seawater Intrusion in Coastal Aquifers-Concepts, Methods and Practices*; Springer: Kluwer, The Netherlands, 1999; pp. 51–71.

6.  Meybeck, M.; Vorosmarty, C.; Schultze, R.; Becker, A. Conclusions: Scaling Relative Responses of Terrestrial Aquatic Systems to Global Changes. In *Vegetation, Water, Humans and the Climate*; Kabat, P., Claussen, M., Dirmeyer, P.A., Gash, J.H.C., Bravo de Guenni, L., Meybeck, M., Pielke, R.A., Vorosmarty, C.J., Hutjes, R.W.A., Lutkemeier, S., Eds.; Springer: Berlin/Heidelberg, Germany, 2003; pp. 455–464.

7.  Werner, A.D.; Bakker, M.; Post, V.E.A.; Vandenboede, A.; Lu, C.H.; Ataie-Ashtiani, B.; Simmons, C.T.; Barry, D.A. Seawater intrusion processes, investigation and management. Recent advances and futures challenges. *Adv. Water Resour.* **2013**, *51*, 3–26. [CrossRef]

8.  Andreasen, D.C.; Fleck, W.B. Use of bromide: Chloride ratios to differentiate potential sources of chloride in a shallow, unconfined aquifer affected by brackish-water intrusion. *Hydrogeol. J.* **1997**, *5*, 17–26. [CrossRef]

9.  Lin, J.; Snodsmith, J.B.; Zheng, C.; Wu, J. A modeling study of seawater intrusion in Alabama Gulf Coast, USA. *Environ. Geol.* **2009**, *57*, 119–130. [CrossRef]

10.  Meisler, H.; Leahy, P.P.; Knobel, LL. *The Effect of Eustatic Sea-Level Changes on Saltwater-Freshwater Relations in the Northern Atlantic Coastal Plain*; U.S. Geological Survey Water-Supply Paper; U.S. Geological Survey: Reston, VA, USA, 1985; Volume 2255, p. 28.

11.  Stringfield, V.T.; LeGrand, H.E. Relation of sea water to fresh water in carbonate rocks in coastal areas, with special reference to Florida, U.S.A. *J. Hydrol.* **1969**, *9*, 387–404. [CrossRef]

12.  Wicks, C.M.; Herman, J.S. Regional hydrogeochemistry of a modern coastal mixing zone. *Water Resour. Res.* **1996**, *32*, 401–407. [CrossRef]

13.  Wicks, C.M.; Herman, J.S.; Randazzo, A.F.; Jee, J.L. Water-rock interactions in a modern coastal mixing zone. *Geol. Soc. Am. Bull.* **1995**, *107*, 1023–1032. [CrossRef]

14.  Barlow, M.P. *Groundwater in Freshwater-Saltwater Environments of the Atlantic Coast*; U.S. Geological Survey Circular: Reston, GA, USA, 2003; p. 1262.

15.  Langman, J.B.; Ellis, A.S. A multi-isotope ($\delta D$, $\delta^{18}O$, $^{87}Sr/^{86}Sr$, and $\delta^{11}B$) approach for identifying saltwater intrusion and resolving groundwater evolution along the Western Caprock Escarpment of the Southern High Plains, New Mexico. *Appl. Geochem.* **2010**, *25*, 159–174. [CrossRef]

16.  Izbicki, J.A. Chloride Sources in a California Coastal Aquifer, in Peters. In *Ground Water in the Pacific Rim Countries*; Peters, H.J., Ed.; American Society of Civil Engineers: Reston, GA, USA, 1991; pp. 71–77.

17.  Izbicki, J.A. Use of $\delta^{18}O$ and $\delta D$ to define seawater intrusion. In *North American Water and Environmental Congress*; Bathala, C.T., Ed.; IR Div/ASCE; American Society of Civil Engineers: New York, NY, USA, 1996.

18.  Todd, D.K. *Sources of Saline Intrusion in the 400-Foot Aquifer, Castroville Area, California*; Monterey Country Flood Control and Water Conservation District: Salinas, CA, USA, 1989; p. 41.

19.  Vengosh, A.; Gill, J.; Davisson, M.L.; Hudson, G.B. A multi-isotope (B, Sr, O, H, and C) and age dating ($^3$H-$^3$He and $^{14}$C) study of groundwater from Salinas Valley, California: Hydrochemistry, dynamics, and contamination processes. *Water Resour. Res.* **2002**, *38*. [CrossRef]

20.  Cardoso, P. Raline water intrusion in Mexico. In *Water Pollution 2018*, 2nd ed.; Transactions on Ecology and the Environment; WIT Press: Southampton, UK, 1993. [CrossRef]

21.  Bocanegra, E.; Da Silva, G.C.; Custodio, E.; Manzano, M.; Montenegro, S. State of knowledge of coastal aquifer management in South America. *Hydrogeol. J.* **2010**, *18*, 261–267. [CrossRef]

22.  Werner, A.D. A review of seawater intrusion and its management in Australia. *Hydrogeol. J.* **2010**, *18*, 281–285. [CrossRef]

23.  Steyl, G.; Dennis, I. Review of coastal-area aquifers in Africa. *Hydrogeol. J.* **2010**, *18*, 217–225. [CrossRef]

24.  Chaouni Alia, A.; El Halimi, N.; Walraevens, K.; Beeuwsaert, E.; De Breuck, W. *Investigation de la Salinisation de la Plaine de Bou-Areg (Maroc Nord-Oriental)*; Freshwater Contamination; International Association of Hydrological Sciences—IAHS Publication: London, UK, 1997; Volume 243, pp. 211–220.

25.  El Halimi, N.; Chaouni Alia, A.; Beeuwsaert, E.; Walraevens, K. Hydrogeological and Geophysical Investigation for Characterizing the Groundwater Reservoir in Saidia Plain (north-eastern Morocco). In *Development of Water Resource Management Tools for Problems of Seawater Intrusion and Contamination of Fresh-Water Resources in Coastal Aquifers*; Walraevens, K., Ed.; Ghent University: Ghent, Belgium, 2000; pp. 67–75, ISBN 90-76878-01-3.

26. Lamrini, A.; Beeuwsaert, E.; Walraevens, K. Contribution to the characterization of the Martil Coastal Aquifer System. In *Development of Water Resource Management Tools for Problems of Seawater Intrusion and Contamination of Fresh-Water Resources in Coastal Aquifers*; Walraevens, K., Ed.; Ghent University: Ghent, Belgium, 2000; pp. 76–81, ISBN 90-76878-01-3.

27. Tarhouni, J.; Jemai, S.; Walraevens, K.; Rekaya, M. Caractérisation de l'aquifère côtier de Korba au Cap Bon (Tunisie). In *Development of Water Resource Management Tools for Problems of Seawater Intrusion and Contamination of Fresh-Water Resources in Coastal Aquifers*; Walraevens, K., Ed.; Ghent University: Ghent, Belgium, 2000; pp. 11–27, ISBN 90-76878-01-3.

28. Imerzoukene, S.; Walraevens, K.; Feyen, J. Salinization of the coastal and eastern zones of the alluvial and unconfined aquifer of the Mitidja Plain (Algeria). In Proceedings of the 13th Salt Water Intrusion Meeting, Cagliari, Italy, 5–4 June 1994; pp. 163–175.

29. Van Camp, M.; Mjemah, I.C.; Alfarrah, N.; Walraevens, K. Modeling approaches and strategies for data-scarce aquifers: Example of the Dar es Salaam aquifer in Tanzania. *Hydrogeol. J.* **2013**, *21*, 341–356. [CrossRef]

30. Walraevens, K.; Mjemah, I.C.; Mtoni, Y.; Van Camp, M. Sources of salinity and urban pollution in the Quaternary sand aquifers of Dar es Salaam, Tanzania. *J. Afr. Earth Sci.* **2015**, *102*, 149–165. [CrossRef]

31. Oude Essink, G.H.P. Saltwater intrusion in 3D larg-scale aquifers a Duch case. *Phys. Chem. Earth* **2001**, *26*, 337–344. [CrossRef]

32. Vandenbohede, A.; Walraevens, K.; De Breuck, W. What does the interface on the fresh-saltwater distribution map of the Belgian coastal plain represent? *Geol. Belg.* **2015**, *18/1*, 31–36.

33. Alcala, F.J.; Custodio, E. Using the Cl/Br ratio as a tracer to identify the origin of salinity in aquifers in Spain and Portugal. *J. Hydrol.* **2008**, *359*, 189–207. [CrossRef]

34. Custodio, E. Coastal aquifers of Europe: An overview. *Hydrogeol. J.* **2010**, *18*, 269–280. [CrossRef]

35. De Montety, V.; Radakovitch, O.; Vallet-Coulomb, C.; Blavoux, B.; Hermitte, D.; Valles, V. Origin of groundwater salinity and hydrochemical processes in a confined coastal aquifer: Case of the Rhône delta (Southern France). *Appl. Geochem.* **2008**, *23*, 2337–2349. [CrossRef]

36. UN—United Nations. *Water for People, Water for Life*; UN World Development Report (WWDR); UN: New York, NY, USA, 2003.

37. WHO—World Health Organization. *Guidelines for Drinking-Water Quality: First Addendum to Third Edition, Volume 1 Recommendations*; WHO: Geneva, Switzerland, 2006.

38. Ben-Asher, J.; Beltrao, J.; Costa, M. Modelling the effect of sea water intrusion on ground water salinity in agricultural areas in Israel, Portugal, Spain and Turkey. In Proceedings of the International Symposium on Techniques to Control Salination for Horticultural Productivity, Antalya, Turkey, 7–10 November 2000.

39. Edmunds, W.M.; Milne, C.J. *Palaeowaters in Coastal Europe: Evolution of Groundwater Since the Late Pleistocene*; Geological Society of London: London, UK, 2001.

40. Gordu, F.; Motz, L.H.; Yurtal, R. Simulation of Seawater Intrusion in the Goksu Delta at Silifke, Turkey. In Proceedings of the First International Conference on Saltwater Intrusion and Coastal Aquifers Monitoring, Modeling, and Management, Essaouira, Morocco, 23–25 April 2001.

41. Yazicigil, H.; Ekmekci, M. Perspectives on Turkish ground water resources. *Ground Water* **2003**, *41*, 290–291. [CrossRef] [PubMed]

42. Karahanoglu, N.; Doyuran, V. Finite element simulation of seawater intrusion into a quarry-site coastal aquifer, Kocaeli-Darica, Turkey. *Environ. Geol.* **2003**, *44*, 456–466. [CrossRef]

43. Peters, H.J. *Ground Water in the Pacific Rim Countries*; IR Div/ASCE; American Society of Civil Engineers: New York, NY, USA, 1991; pp. 71–77.

44. Demirel, Z. The history and evaluation of saltwater intrusion into a coastal aquifer in Mersin, Turkey. *J. Environ. Manag.* **2004**, *70*, 275–282. [CrossRef] [PubMed]

45. Camur, M.Z.; Yazicigil, H. Effects of the planned Ephesus recreational canal on freshwater-seawater interface in the Selcuk sub-basin, Izmir. *Environ. Geol.* **2005**, *48*, 229–237. [CrossRef]

46. Alfarrah, N.; Hweesh, A.; Van Camp, M.; Walraevens, K. Groundwater flow and chemistry of the oases of Al Wahat, NE Libya. *Environ. Earth Sci.* **2016**, *75*, 1–24. [CrossRef]

47. Gaaloul, N.; Pliakas, F.; Kallioras, A.; Marinos, P. Seawater intrusion in Mediterranean porous coastal aquifers: Cases from Tunisia and Greece. In Proceedings of the 8th International Hydrogeological Congress of Greece, Athens, Greece, 8–10 October 2008; pp. 281–290.

48. World Bank. *People's Democratic Republic of Algeria—A Public Expenditure Review. Assuring High Quality Public Investment*; Report No. 36270; World Bank: Washington, DC, USA, 2007.

49. Djabri, L.; Laouar, R.; Hani, A.; Mania, J.; Mudry, J. The Origin of Water Salinity on the Annaba Coast (NE Algeria). In Proceedings of the Symposium HSO2a—IUGG2003, Sapporo, Japan, 30 June–11 July 2003; International Association of Hydrological Sciences—IAHS Publication: London, UK, 2003; Volume 280, pp. 229–235.

50. Pulido-Bosch, A.; Tahiri, A.; Vallejos, A. Hydrogeochemical characteristics of processes in the Temara Aquifer in Northwestern Morocco. *Water Air Soil Pollut.* **1999**, *114*, 323–337. [CrossRef]

51. Sherif, M.M. Nile Delta aquifer in Egypt. In *Seawater Intrusion in Coastal Aquifers—Concepts, Methods and Practices*; Bear, J., Cheng, A., Sorek, S., Ouazar, D., Herrera, A., Eds.; Theory and Application of Transport in Porous Media Book Series; Kluwer Academic Publishers: Dordrecht, The Netherlands, 1999; pp. 559–590.

52. ISARM. *Internationally Shared (Transboundary) Aquifer Resources Management*; IHP-VI, IHP Non Serial Publications in Hydrology November, Paris: UNESCO 2001; Internationally Shared (Transboundary) Aquifer Resources Management: Delft, The Netherlands, 2001.

53. Khouri, J. Water resources of the Zarka River basin, Jordan. In *Water Resources Management and Desertification Problems and Challenges*; World Meteorological Organization (WMO): Geneva, Switzerland, 1996; pp. 84–96.

54. Alfarrah, N. Hydrogeological and Hydrogeochemical Investigation of the Coastal Area of Jifarah Plain, NW Libya. Ph.D. Thesis, Laboratory of Applied Geology and Hydrogeology, Ghent University, Ghent, Belgium, 2011.

55. Alfarrah, N.; Berhane, G.; Bakundukize, C.; Walraevens, K. Degradation of groundwater quality in coastal aquifer of Sabratah area, NW Libya. *Environ. Earth Sci.* **2017**, *76*, 664. [CrossRef]

56. Alfarrah, N.; Van Camp, M.; Walraevens, K. Deducing transmissivity from specific capacity in the heterogeneous upper aquifer system of Jifarah Plain, NW-Libya. *J. Afr. Earth Sci.* **2013**, *85*, 12–21. [CrossRef]

57. Gefli. *Soil and Water Resources Survey for Hydro-Agricultural Development, Western Zone*; Unpublished Report; Ground Water Authority: Tripoli, Libya, 1972.

58. Libyan Industrial Research Centre (IRC). *Geological Map of Jifarah Plain*, 1st ed.; IRC: Tajura, Libya, 1975.

59. Kruseman, G.P. *Evaluation of Water Resources of the Gefara Plain*; Unpublished Report; SDWR: Tripoli, Libya, 1977.

60. Krummenacher, R. *Gefara Plain Water Management Plan Project*; Report on the Groundwater Resources of the Gefara Plain; Unpublished Report; 110p and 4 Annexes; GWA: Tripoli, Libya, 1982.

61. APHA (American Public Health Association). *Standard Methods for the Examination of Water and Wastewater*; Greenberg, A.E., Trussell, R.R., Clesceri, L.S., Eds.; APHA: Washington, DC, USA, 1985.

62. Parkhurst, D.L.; Appelo, C.A.J. *User's Guide to PHREEQC (Version 2)-A Computer Program for Speciation, Batch-Reaction, One-Dimensional Transport, and Inverse Geochemical Calculations*; U.S. Geological Survey Water-Resources Investigations Report 99-4259; U.S. Geological Survey: Reston, VA, USA, 1999; p. 312.

63. Fidelibus, M.D.; Tulipano, L. Regional flow of intruding seawater in the carbonate aquifers of Apulia (Southern Italy). In Proceedings of the 14th Salt Water Intrusion Meeting, Malmö, Sweden, 17–21 June 1996; Geological Survey of Sweden: Uppsala, Sweden, 1996.

64. Appelo, C.A.J. Cation and proton exchange, pH variations, and carbonate reactions in a freshening aquifer. *Water Resour. Res.* **1994**, *30*, 2793–2805. [CrossRef]

65. Stuyfzand, P.J. A new hydrogeochemical classification of water types: Principles and application to the coastal dunes aquifer system of the Netherlands. In Proceedings of the 9th SWIM, Delft, The Netherlands, 12–16 May 1986; pp. 641–656.

66. Stuyfzand, P.J. Hydrochemistry and Hydrology of the Coastal Dune Area of the Western Netherlands. Ph.D. Thesis, Free University (VU), Amsterdam, The Netherlands, 1993; p. 366.

67. Mollema, P.N.; Antonellini, M.; Dinelli, E.; Gabbianelli, G.; Greggio, N.; Stuyfzand, P.J. Hydrochemical and physical processes influencing salinization and freshening in Mediterranean low-lying coastal environments. *Appl. Geochem.* **2013**, *34*, 207–221. [CrossRef]

68. Mollema, P.N.; Antonellini, M.; Stuyfzand, P.J.; Juhasz-Holterman, M.H.A.; Van Diepenbeek, P.M.J.A. Metal accumulation in an artificially recharged gravel pit lake used for drinking water supply. *J. Geochem. Explor.* **2015**, *150*, 35–51. [CrossRef]

69. Mollema, P.N. Water and Chemical Budgets of Gravel Pit Lakes: Case Studies of Fluvial Gravel Pit Lakes along the Meuse River (The Netherlands) and Coastal Gravel Pit Lakes along the Adriatric Sea (Ravenna, Italy). Ph.D. Thesis, Technische Universiteit Delft, Delft, The Netherlands, 2016.

70. WHO—World Health Organization. *Guidelines for Drinking-Water Quality (Electronic Resource): Incorporating 1st and 2nd Addenda, Volume 1, Recommendations*, 3rd ed.; WHO Library Cataloguing-in Publication Data; WHO: Geneva, Switzerland, 2008; p. 668.

71. Mercado, A. The use of hydrogeochemical patterns in carbonate sand and sandstone aquifers to identify intrusion and flushing of saline waters. *Groundwater* **1985**, *23*, 635–664. [CrossRef]

72. El Moujabber, M.; Bou Samra, B.; Darwish, T.; Atallah, T. Comparison of different indicators for groundwater contamination by seawater intrusion on the Lebanese coast. *Water Resour. Manag.* **2006**, *20*, 161–180. [CrossRef]

73. Walraevens, K.; Van Camp, M. Advances in understanding natural groundwater quality controls in coastal aquifers. In Proceedings of the 18th Salt Water Intrusion Meeting (SWIM), Cartagena, Spain, 31 May–3 June 2004; pp. 451–460.

74. Vengosh, A.; Starinsky, A.; Melloul, A.; Fink, M.; Erlich, S. *Salinization of the Coastal Aquifer Water by Ca-Chloride Solutions at the Interface Zone, Along the Coastal Plain of Israel*; Hydrological Service: Jerusalem, Israel, 1991.

75. Jeen, S.W.; Kim, J.M.; KO, K.S.; Yum, B.W.; Chang, H.W. Hydrogeochemical characteristics of groundwater in a Midwestern coastal aquifer system, Korea. *Geosciences* **2001**, *5*, 339–348. [CrossRef]

76. Sanchez-Martos, F.; Pulido-Bosch, A.; Calaforra-Chordi, J.M. Hydrogeochemical processes in an arid region of Europe (Almeria, SE Spain). *Appl. Geochem.* **1999**, *14*, 735–745. [CrossRef]

MDPI

St. Alban-Anlage 66

4052 Basel

Switzerland

Tel. +41 61 683 77 34

Fax +41 61 302 89 18

www.mdpi.com

*Water* Editorial Office

E-mail: water@mdpi.com

www.mdpi.com/journal/water

www.ingramcontent.com/pod-product-compliance
Lightning Source LLC
Chambersburg PA
CBHW051859210326
41597CB00033B/5951